Gaswell Testing

Gaswell Testing

Theory, Practice & Regulation

David A.T. Donohue
President, IHRDC,
Arlington Exploration Company

Turgay Ertekin
Assistant Professor,
The Pennsylvania State University

with the assistance of
Howard B. Bradley

International Human Resources Development Corporation
Boston

ISBN-13: 978-0-934634-12-0   e-ISBN-13: 978-94-011-7449-7
DOI: 10.1007/978-94-011-7449-7

Library of Congress Catalog Card Number: 81–80726

Table of Contents

Acknowledgments

The authors of this book wish to thank the following individuals for the active support and encouragement they provided during its several stages of completion:

Howard B. Bradley of Mobil Exploration and Producing Services, Inc. in Dallas who reviewed the full manuscript including the tables, figures and equations, offered continued support and criticism, and whose personal interest in both video production and the book led to its satisfactory completion;

Aziz S. Odeh and *William Prachner* of Mobil Oil Corporation, Dallas, for the valuable suggestions they offered during the planning and working stages of the book and video production;

Bruce Jankura of Mobil Oil Corporation, Lake Charles, Louisiana for advice during planning and for his willingness to appear in the video production;

Grant Johnson of Mobil Canada Ltd., Edmonton and *R.G. Justice* of Mobil Oil Corporation, Denver for their advice and recommendations during writing;

Derek F. Harvey of Mobil Oil Corporation, New York for his careful helmsmanship; and especially to *Michael Hays* and *Gail Smith* of IHRDC for editorial assistance, review, and final publication.

Preface

This book accompanies a videotape program of the same name. The combined videotape and book, referred to as a module of instruction, was one of three prepared by IHRDC on a joint basis with Mobil Oil Corporation during 1980. The three modules, one each in geology, geophysics, and petroleum engineering, were produced to determine whether this medium of instruction would provide an effective way of teaching recent graduates and those individuals changing specialties, "what they need to know, when they need to know it." A great deal was learned during the pilot stages. Properly designed and properly used, video-assisted instruction was found to be effective, efficient, and convenient.

With the confidence that this instructional medium provides one way for the international petroleum industry to train young graduates in exploration and production, IHRDC sought financial and advisory support from a limited number of companies to undertake the development of the *Basic Technical Video Library for the E&P Specialist*. To date the following companies have agreed to serve as Sponsors: Mobil, AGIP, ARAMCO, Cities Services, Dome Petroleum Ltd., Gulf, Phillips, Standard Oil of California/Chevron, and Texaco.

Work on the Library began in July 1981. With an accelerated production schedule of 24 modules per year and the continued support of the Sponsors, the Library should be completed in about five years.

Introduction

Gas wells are tested to determine their ability to produce gas under various conditions of surface or sandface (sandface pressure as used here is the same as the bottomhole tubing pressure) and reservoir pressures. The controlling surface pressure will depend on the operating pressure of the pipeline to which the gas is delivered, the characteristics of any installed compression, the pressure losses that occur in gathering lines, and the choke size selected to control the production rate. Also, pressure losses that occur in the tubing must be added. The reservoir pressure depends on the extent of reservoir depletion and on any pressure support that might be provided by water encroachment. A typical flow system is shown in figure I.1.

Several general conclusions may be drawn from figure I.1:

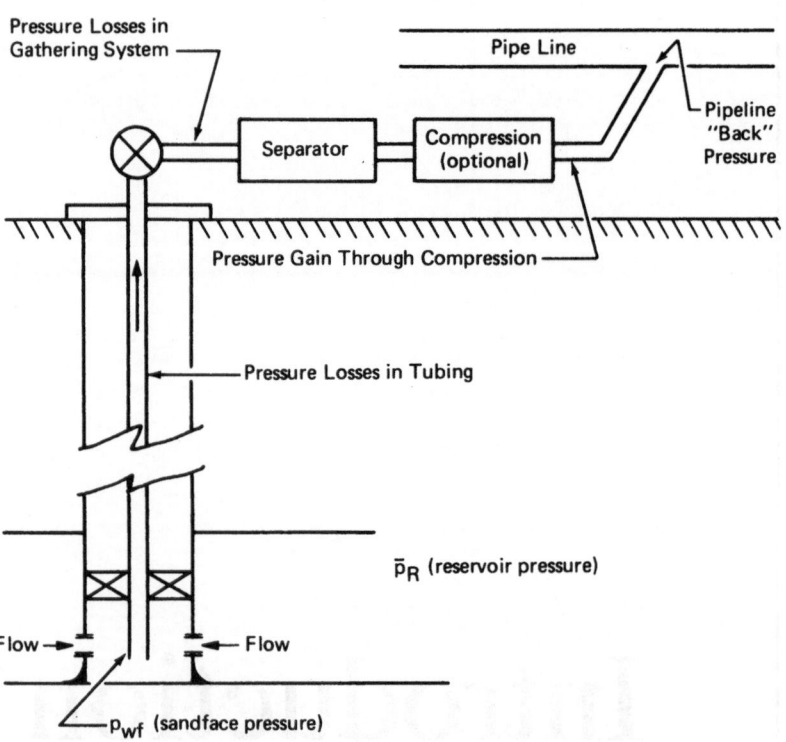

Figure I.1
Schematic of a Gaswell System.

2   Gaswell Testing

(1) Maximum delivery will occur when the reservoir pressure, $\bar{p}_R$, is a maximum and the sandface pressure, p_{wf}, is a minimum.

(2) A high pipeline pressure, without installed compression, will reduce deliverability.

(3) Pressure losses in the gathering system and wellbore will reduce deliverability.

(4) Installation of compression will increase deliverability.

(5) For a given sandface pressure, deliverability will decrease as the reservoir pressure is reduced through depletion.

(6) The formation characteristics of wells are different; therefore, each must be tested to determine its unique ability to produce.

Because the capacity of a well to produce into a wellbore is a unique characteristic of the well (its subsurface and surface flow system), field testing procedures have been developed that allow the engineer to predict the manner in which a well will produce over its reserve's life under the various operating conditions that may be imposed. The results of these tests are used by the engineer to:

- Determine whether a well will be commercial.
- Satisfy the regulations of state agencies.
- Determine allowable gas production rates.
- Design processing plants and pipeline extensions.
- Serve as a basis for gas sales contracts.
- Support deliverability and compression studies.
- Determine well spacing and field development programs.
- Plan cycling programs.
- Determine the need for stimulation.
- Assist in the identification of reservoir boundaries such as faults, pinchouts, and water tables.

Accordingly, it is important that the engineer understand fully the background, procedures, and interpretation of gaswell tests.

The purpose of this book is to cover the principal aspects of gaswell testing. The chapters follow a topical sequence: chapters 1 and 2 are devoted to the theoretical fundamentals of testing; chapter 3 to test design considerations; chapter 4 to test procedures and state/province regulations; chapter 5 to the procedures followed and interpretation made for a specific gaswell test in Louisiana, and chapter 6 to exercises. Mathematical developments, correlations, sample reporting forms, and similar material are given as appendixes.

Chapter 1

Conventional & Isochronal Tests

Deliverability tests on gas wells have been used for many years to determine flow capacity. In early times gas wells were rated by opening the well fully to the atmosphere and determining its open flow capacity. In order to prevent wastage of gas and possible formation damage, open flow potential (capacity) tests of gas wells were soon determined by flowing the wells against a particular pipeline backpressure. Since a backpressure is maintained at the surface during these tests, they are commonly called backpressure tests.

1.1 Derivation of Fundamental Flow Equation

In order to understand the relationships that exist when gas flows through a formation into a wellbore, certain mathematical developments are necessary. We start with the derivation of Darcy's law for the radial flow of gases. Consider radial flow toward a vertical wellbore of radius r_w in a horizontal reservoir of uniform thickness and permeability as shown in figure 1.1. The radius at the external boundary is r_e and the pressure is p_e.

If the flow rate of the gas at any radius, r, is q_r it can be shown that

$$q_{rw} \left(\frac{r_e^2 - r^2}{r_e^2} \right) = q_{rw} \left(1 - \frac{r^2}{r_e^2} \right) \qquad (1.1)$$

where $q_{rw} = \dfrac{2\pi rhk}{\mu_g z}$ at $r = r_w$, i.e., the flow rate at the we

By Darcy's Law:

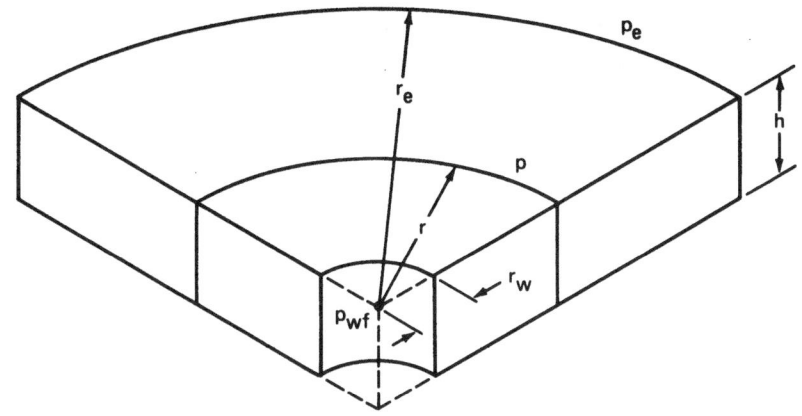

Figure 1.1
Schematic Representation and Nomenclature for a Radial Flow System.

$$q_r = q_{rw} \left(\frac{1}{r} - \frac{r}{r_e^2} \right) dr = \frac{2\pi hkr}{\mu_g z} \frac{dp}{dr} \quad (1.2)$$

at standard conditions

$$q_r = \frac{q_{sc} \, p_{sc} \, z \, T_R}{T_{sc} \, p} = q_{sc} \, G/P \quad (1.3)$$

and

$$q_{sc} G \int_{r_w}^{r_e} \left(\frac{1}{r} - \frac{r}{r_e^2} \right) dr = \frac{2\pi hk}{\mu_g z} \int_{P_{wf}}^{P_e} p \, dp \quad (1.4)$$

Integrating gives:

$$q_{sc} C \left(\ln \frac{r_e}{r_w} - 1/2 \right) = \frac{2\pi kh}{\bar{\mu}_g \bar{z}} (p_e^2 - p_{wf}^2) \text{ or}$$

$$q_{sc} = \frac{0.703 \times 10^{-6} \, kh \, (p_e^2 - p_{wf}^2)}{\bar{\mu}_g \, T_R \, \bar{z} \left(\ln \left(\frac{r_e}{r_w} \right) - 1/2 \right)} \quad (1.5)$$

The equation (1.4) can be integrated from p_w to p to obtain the pressure distribution as a function of r. The average reservoir pressure,

$$\bar{p}_R = \frac{2}{r_e^{\,2}} \int_{r_w}^{r_e} p_{(r)}^2 \; r dr \qquad (1.6)$$

Substituting for $p_{(r)}^2$ from the previous step, integrating and rearranging gives:

$$q_{sc} = \frac{0.703 \times 10^{-6}\, kh\,(\bar{p}_R^{\,2} - p_{wf}^{\,2})}{\bar{\mu}_g\, T_R\, \bar{z}\left(\ln\left(\dfrac{r_w}{r_e}\right) - 0.75\right)} \qquad (1.7)$$

In equations (1.5) and (1.7), the pressure-dependent parameters of viscosity μ_g and compressibility factor z are assumed to be evaluated at the average reservoir pressure $\bar{p}_R$ and therefore treated as constants.

Now let

$$C = \frac{0.703 \times 10^{-6}\, kh}{\bar{\mu}_g T_R\; \bar{z}\left(\ln\left(\dfrac{r_e}{r_w}\right) - 0.75\right)} \qquad (1.8)$$

Then

$$q_{sc} = C\,(\bar{p}_R^{\,2} - p_{wf}^{\,2}) \qquad (1.9)$$

where

$\bar{p}_R$ = average reservoir pressure, psia
p_{wf} = flowing bottomhole pressure at r_w, psia
k = permeability of the porous medium, md
h = formation thickness, ft
T_R = reservoir temperature, °R
$\bar{\mu}_g$ = gas viscosity at average reservoir pressure, cp

$\bar{z}$ = gas compressibility factor at average reservoir pressure, fraction

r_e = external radius, ft

r_w = well radius, ft

q_{sc} = production rate, MMSCF/D

C = performance constant for the well

Since the ratio r_e/r_w in equation (1.7) and others to follow is usually very large, the -0.75 value is sometimes omitted in deriving the equation. Taking the logarithm of both sides of equation (1.9)

$$\log q_{sc} = \log C + \log (\bar{p}_R^2 - p_{wf}^2) \qquad (1.10)$$

This equation indicates that for Darcy-type laminar flow, the log-log plot of q_{sc} versus $(\bar{p}_R^2 - p_{wf}^2)$ is a straight line whose slope is 45°. Because of deviations from ideal flow behavior, the actual plots may be nonlinear and their slopes less than 1.00. Rawlins and Schellhardt,[1] drawing upon empirical observations, modified equation (1.9) by adding the exponent n to account for the observed deviation from the ideal flow behavior. The deviation is due to several factors, the most important being the effect of turbulence. Thus, equation (1.9) takes the following form.

$$q_{sc} = C(\bar{p}_R^2 - p_{wf}^2)^n \qquad (1.11)$$

The range of values for the exponent n in the above equation varies between 1.0 and 0.5, approaching 1.0 and 0.5 at complete laminar and turbulent flow conditions, respectively.

The logarithm of both sides of equation (1.11) yields,

$$\log q_{sc} = \log C + n \log (\bar{p}_R^2 - p_{wf}^2) \qquad (1.12)$$

From the definition of performance factor, C, it is obvious that this factor depends on gas properties, such as viscosity, compressibility factor, and reservoir properties, such as permeability, temperature, formation thickness, external boundary radius, and wellbore radius.

1.2 Conventional Test

Looking at equation (1.12) it is seen that a log-log plot of $(\bar{p}_R^2 - p_{wf}^2)$ versus q_{sc} should give a straight line whose slope is $1/n$. In order to obtain a straight line we need at least two points through which the line may be drawn. In general practice it is suggested that four points be obtained to minimize error.

To perform a conventional test, the static reservoir pressure, $\bar{p}_R$, is determined by shutting in the well for a period of time until its stabilized pressure is known. A flow rate, q_{sc}, is then selected, and the well is flowed until its flowing pressure stabilizes. In many areas, stabilization is defined in terms of percentage change per unit of time. The stabilized flowing pressure, p_{wf}, is recorded. This procedure is repeated four times by sequentially decreasing or, preferably, by increasing the flow rate and measuring the corresponding stabilized flowing pressure each time. The flow rate and pressure histories for a conventional test are shown schematically in figure 1.2.

Now, we can plot the point $(\bar{p}_R^2 - p_{wf}^2)$ versus q_{sc} for each measured rate and draw a best-fit line through the points as shown in figure 1.3.

The straight line of figure 1.3 has a slope $1/n$ and is known as the backpressure line or inflow performance line. From this straight line and equation (1.12) one can determine the absolute open flow potential or AOF, the theoretical flow rate that would occur if the sandface pressure was

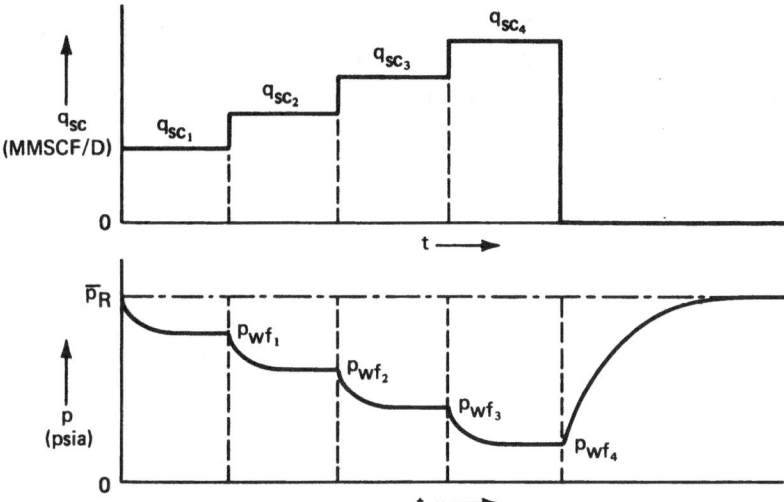

Figure 1.2
Flow Rate and Pressure History of a Typical Conventional Test.

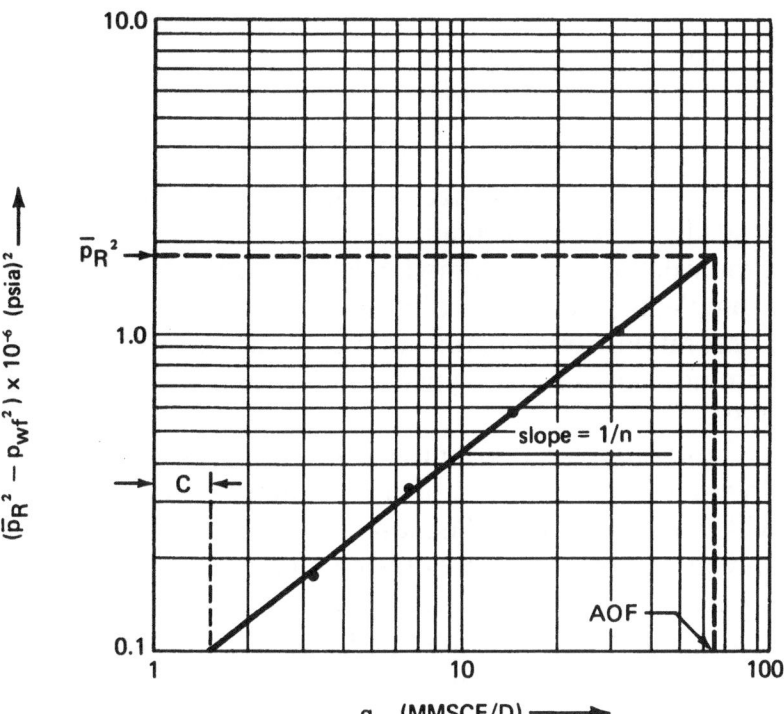

Figure 1.3
Plot for Conventional Well Test Example.

Conventional & Isochronal Tests **11**

reduced to zero. Having obtained values of n and C, equation (1.11) may then be used to calculate the deliverability (inflow rate) of a well for any given set of sandface and reservoir pressures.

Example 1.1
Deliverability Calculation for a
Conventional Test

The following example illustrates the application of equation (1.12).

Suppose that we know that $C = 0.0037$ (for q_{sc} in MMSCF/D) and $n = 0.93$, what is the flow rate when $\bar{p}_R = 3000$ psia and $p_{wf} = 1850$ psia?

$$q_{sc} = C(\bar{p}_R^2 - p_{wf}^2)^n$$
$$q_{sc} = 0.0037[(3000)^2 - (1850)^2]^{0.93}$$
$$q_{sc} = 6.96 \text{ MMSCF/D}$$

What is the AOF?

$$AOF = 0.037[(3000)^2 - (0^2)]^{0.93}$$
$$AOF = 10.86 \text{ MMSCF/D}$$

It should be recognized that only a single flow-rate test is needed in order to obtain the backpressure flow equation if the value for n is assumed or known. This is the case, for example, for the single-point test allowed by the Oklahoma Corporation Commission—in that particular situation, the commission allows the operator to assume a value for n of 0.85.

Although the conventional test requires four constant flow rates, it is highly unlikely in practice, that a flow rate will remain constant. Because of wellbore storage effects and because sandface pressure changes occur, the flow rate during each flow period will decrease with time as shown in figure 1.4. Winestock and Colpitts,[2] to account for this variation in rate, suggested that the instantaneous flow rate (say at the end of a flow period) be used rather than an

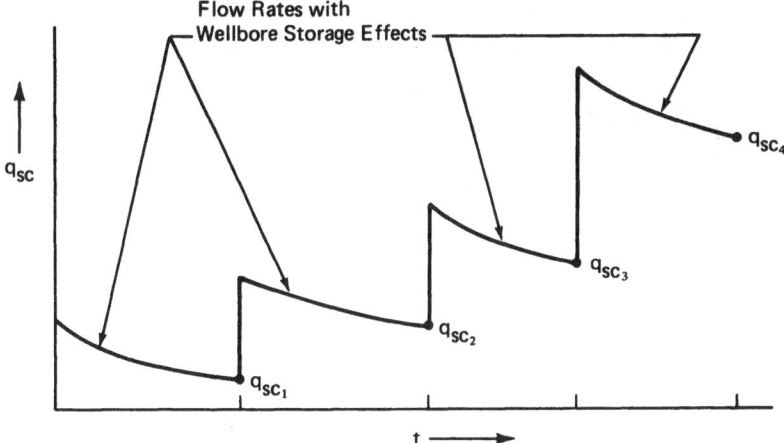

Figure 1.4
Schematic Showing Actual Flow Rate during Test.

average flow rate. In order to incorporate variable rate flow tests, they suggest the use of normalized equations.

1.3
Isochronal Test

Conventional gaswell testing and the interpretation of its data are relatively simple. Although the test has been considered the basic standard for many years, it has certain drawbacks. The difficulty arises when the reservoir permeability is low, or flaring is to be minimized. In this type of reservoir a properly stabilized, conventional deliverability test may not be conducted in a reasonable period of time. In other words, the time required to obtain stabilized flow conditions may be very long.

Cullender[3] proposed the isochronal gaswell test, a test in which a well is shut in long enough before each test-flow period so that each flow will begin with the same pressure distribution in the reservoir. The basic principle behind isochronal testing is that the effective drainage radius, a function of the duration of flow, is the same for each measured data point. Thus, separate flow tests, run for the same length of time, will affect the same radius of drainage. Therefore, each performance curve (or line on a log-log plot) will be obtained for a fixed radius of drainage. Consequently, each of the isochronal curves for a well will have the same slope but the value of the performance constant, C, will depend on the duration of flow (or radius of drainage). An accurate value for C will be obtained with a flow test that reflects the behavior of a large radius of drainage (often a long time).

The procedure normally followed in this test is outlined below (also see figure 1.5).

(1) The well is initially shut in until the static pressure, $\bar{p}_R$, stabilizes.
(2) The well is opened to production at the first rate, q_{sc}, flowed for a predetermined and fixed period of time, and the bottomhole pressure is measured.
(3) Steps 1 and 2 are repeated two or three additional

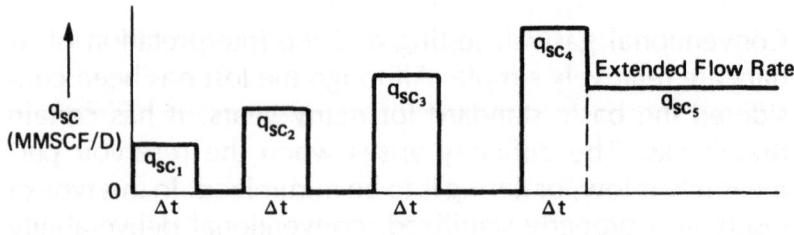

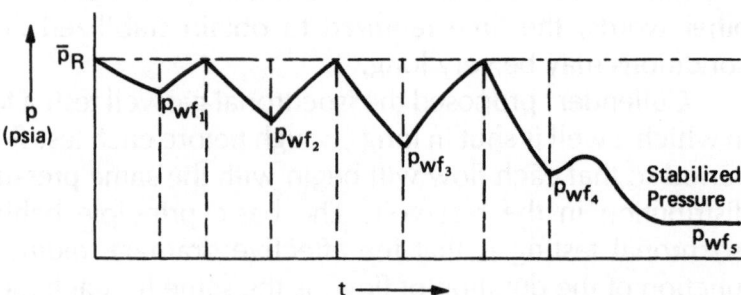

Figure 1.5
Flow Rate and Pressure History of an Isochronal Test.

14 Gaswell Testing

times at different flow rates. The duration of each flow period should be equal and the pressure after each flow period must build up to its original static value. The duration of buildup, then, will not normally be equal to that for the flowing periods.

(4) Finally, after the last flow period, one flow test is conducted for a time period long enough to attain stabilized flow conditions. This period is usually called an extended or stabilized flow period. The rate of flow during this last flow test need not be equal to that of the fourth isochronal rate, and, in fact, is often less. The time required to reach stabilized pressure will depend on the properties of the formation and the fluids.

The isochronal flow data that are obtained following the above outlined procedure are analyzed as follows:

(1) The three or four isochronal points* are plotted on log-log paper as shown in figure 1.6.

(2) The best straight line is drawn through the isochronal points.

(3) The value of the exponent n is obtained from the slope of this line. Remember the inverse of the slope will give the value of n. The value of n, obtained from the isochronal test, should be equal to that obtained by the conventional test; however, as explained earlier, the position of the line and, thus the value of C is not correct.

(4) The stabilized deliverability line (line with the correct value of C) is obtained merely by plotting the stabilized point representing the extended flow rate and stabilized flowing sandface pressure, and drawing a line through it parallel to the plotted line.

Any set of data at equal times can be used; not just the final points.

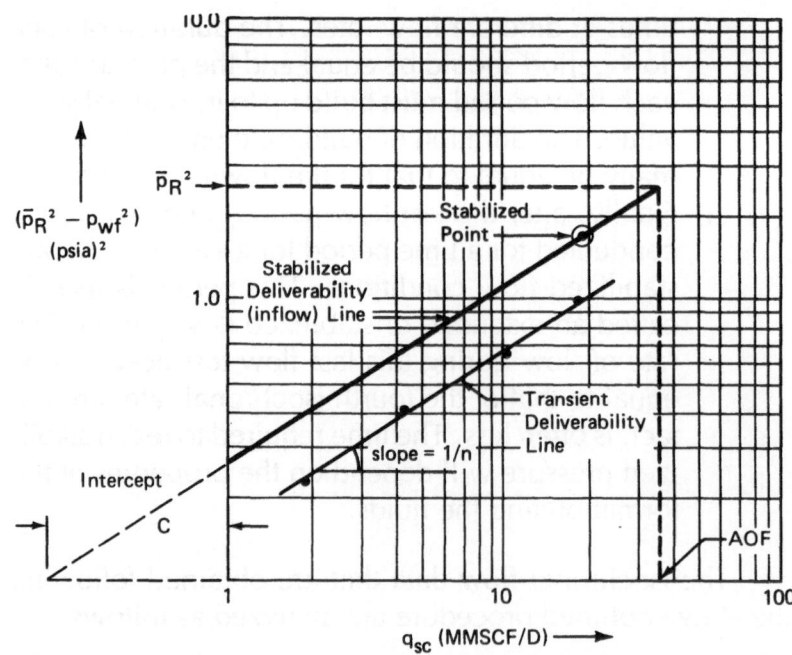

Figure 1.6
Log-log Plot of a Typical Isochronal Well Test.

(5) Provided there are no significant turbulent flow pressure losses at high rates (see section 2.6 for correction for turbulence), the AOF or the flow rate obtained against any sandface backpressure can be easily read from the stabilized inflow performance line.

Example 1.2
Deliverability Calculations for an Isochronal Test

Analyze the following isochronal well test data.

Duration of Flow Test (hours)	Sandface Pressure (psia)	Flow Rate (MMSCF/D)	Shut-in BHP (psia)
Shut-in	2200	0	2200
6	1892	2.8	2200
6	1782	3.4	2200
6	1647	4.8	2200
6	1511	5.4	2200

Afterwards, the well continued to produce at 6 MMSCF/D and reached a stabilized flowing sandface pressure of 1180 psia.

Solution

The following table is prepared:

p_{wf}(psia)	q_{sc} (MMSCF/D)	p_{wf}^2 (psia)^2	$\bar{p}_R^2 - p_{wf}^2$ (psia)^2
2200 = $\bar{p}_R$	0	4.84×10^6	0
1892	2.8	3.58×10^6	1.26×10^6
1782	3.4	3.18×10^6	1.66×10^6
1647	4.8	2.71×10^6	2.13×10^6
1511	5.4	2.28×10^6	2.56×10^6
Stabilized Point			
1180	6.0	1.39×10^6	3.45×10^6

Plot $(\bar{p}_R^2 - p_{wf}^2)$ versus q_{sc} on log-log paper. The slope of the deliverability line is

$$\frac{1}{n} = \frac{\log (5 \times 10^6) - \log (0.43 \times 10^6)}{\log (10^1 \times 10^6) - \log (1 \times 10^6)} = 1.07$$

$$n = 0.94; \quad C = \frac{q_{sc}}{(\bar{p}_R^2 - p_{wf}^2)^n} = \frac{6}{(3.45 \times 10^6)^{0.94}}$$

$$C = 4.29 \times 10^{-6}$$

The AOF from the plot is 8.0 MMSCF/D. See example 1.2 figure. By calculation it is:

$$AOF = 4.29 \times 10^{-6} \, [(2200)^2 - (0)^2]^{0.94} = 8.25 \text{ MMSCF/D}$$

1.4
Modified
Isochronal Test

Compared to the conventional test, the isochronal test should save a considerable volume of gas from being flared

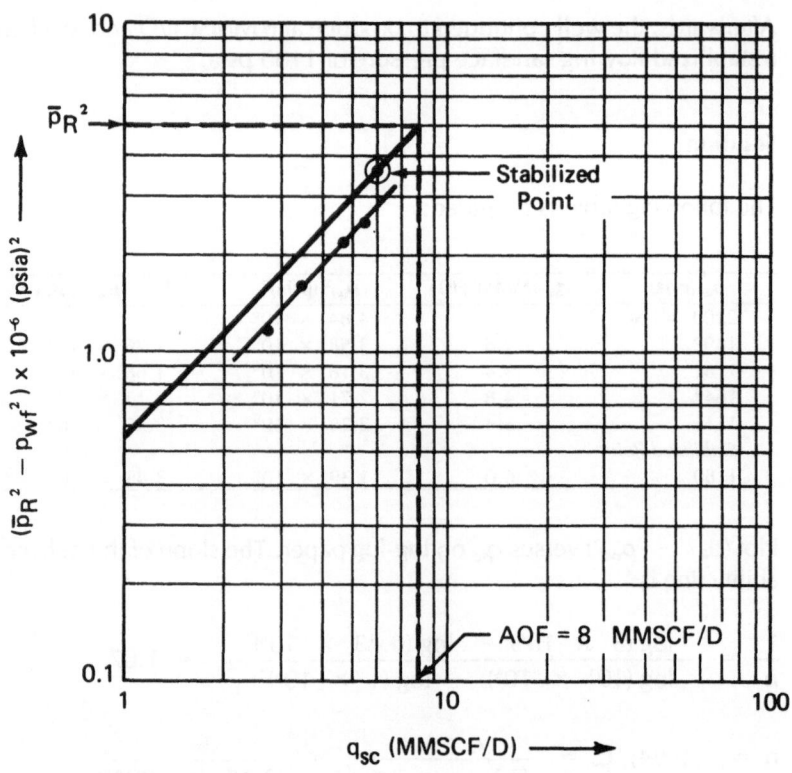

Example 1.2
Plot of Data for Example 1.2.

into the atmosphere. It may also save time if the buildup to static pressure after each flow period is relatively short. This time saving during the flow periods may be considerable in the testing of wells producing from tight gas sands. In extremely tight gas reservoirs, an isochronal test may not always be practical since it is very difficult to attain a completely stabilized static reservoir pressure before the initial flow period and during each subsequent shut-in period.

In 1959, Katz et al.[4] suggested a modification to the isochronal test. They suggested that both the shut-in period and the flow period for each test could be of equal duration provided that the unstabilized shut-in pressure, p_{wR}, at the end of each test be used instead of the static reservoir pressure, $\bar{p}_R$, in calculating the difference of pressure squared for the next flow rate. Figure 1.7 outlines the flow rate and pressure sequence of a typical modified isochronal test.

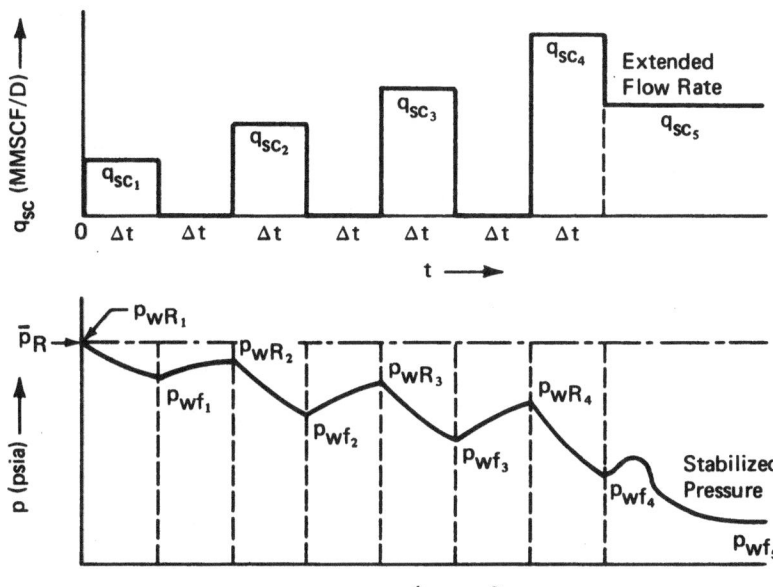

Figure 1.7
Flow Rate and Pressure History for a Typical Modified Isochronal Test.

     The analysis of modified isochronal test data is similar to that shown earlier for the isochronal test. As in the isochronal test, the first four points are plotted on log-log paper as shown in figure 1.8. It is important to remember that the associated unstabilized shut-in pressure is used instead of stabilized static pressure in calculating the difference of pressure-squared for the next flow rate. A best-fit line is drawn through the four points to determine the slope and a second line parallel to the first is drawn through the stabilized point. The values of n, C, and AOF are found in the same manner as outlined in the conventional and isochronal gaswell testing procedures.

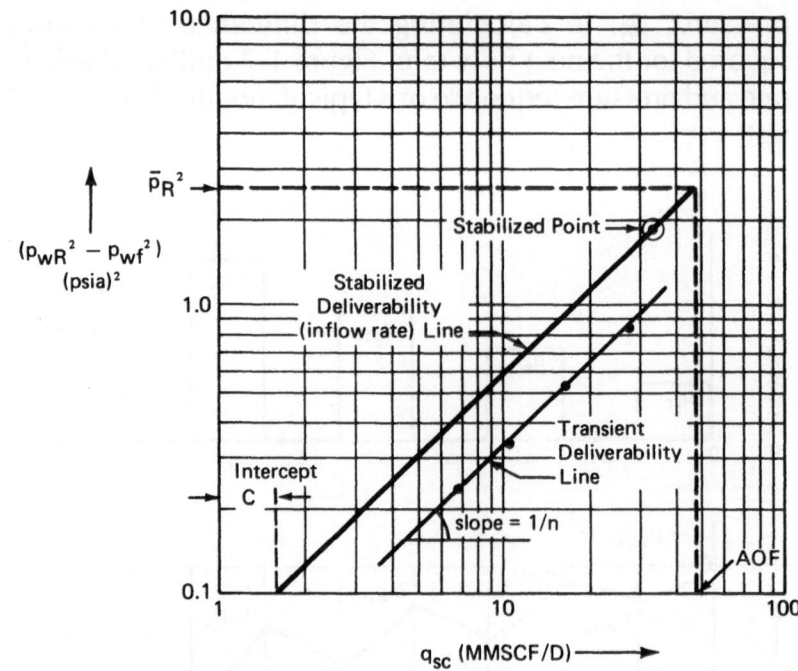

Figure 1.8
Log-log Plot of a Typical Modified Isochronal Gaswell Test.

Example 1.3
Deliverability Calculations for a
Modified Isochronal Test

Analyze the following modified isochronal well test data.

Time of Test (hours)	Sandface Pressure (psia)	Flow Rate (MMSCF/D)	Remarks
14	2000	0	Initial shut-in
10	1842	4.0	Flow 1
10	1982	0	Shut-in
10	1712	6.0	Flow 2
10	1960	0	Shut-in
10	1511	8.0	Flow 3
10	1913	0	Shut-in
10	1306	10.0	Flow 4
26	1072	10.0	Extended flow
68	2000	0	Final shut-in

Solution: Prepare the following table.

p_{wR} (psia)	p_{wf} (psia)	q_{sc} (MMSCF/D)	$p_{wR}^2 - p_{wf}^2$ (psia)^2
2000 = $\bar{p}_R$		0	
	1842	4.0	0.61×10^6
1982		0	
	1712	6.0	0.99×10^6
1960		0	
	1511	8.0	1.56×10^6
1913		0	
	1306	10.0	1.95×10^6
	1072	10.0	2.85×10^6
2000 = $\bar{p}_R$		0	stabilized point

Plot ($p_{wR}^2 - p_{wf}^2$) versus q_{sc} on log-log paper. We can read the slope of the deliverability line as follows:

$$\frac{1}{n} = \frac{\log(2.85 \times 10^6) - \log(0.14 \times 10^6)}{\log(10 \times 10^6) - \log(1 \times 10^6)} = 1.31$$

or $n = 0.76$

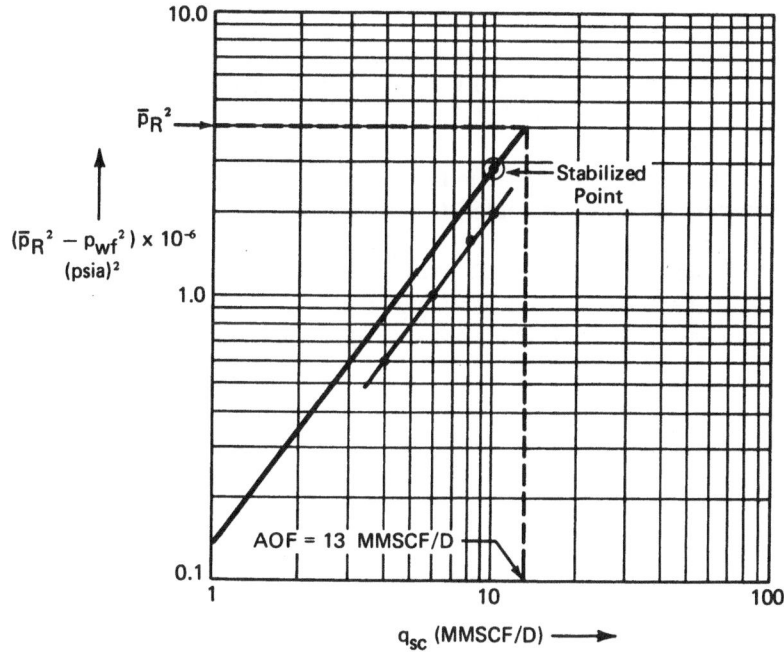

Example 1.3
Plot of Data for Example 1.3.

Then constant C, can be calculated as:

$$C = \frac{q_{sc}}{(p_{wR}^2 - p_{wf}^2)} = \frac{10}{(2.85 \times 10^6)^{0.76}}$$

$$C = 0.000124$$

From the plot shown in figure 1.8 we can read the AOF as being equal to 13 MMSCF/D or alternatively,

$$AOF = C[p_{wR}^2 - (0)^2]^n$$

$$AOF = 0.000124[(2000)^2 - (0)^2]^{0.76}$$

$$AOF = 12.91 \text{ MMSCF/D}$$

which is quite close to our reading from the plot.

References

[1] Rawlins, E.L. and Schellhardt, M.A.: *Backpressure Data on Natural Gas Wells and Their Application to Production Practices*, Monograph 7, U.S. Bureau of Mines, 1936.

[2] Winestock, A.G. and Colpitts, G.P.: "Advances in Estimating Gas Well Deliverability," *J. Can. Pet. Tech.* (July–September, 1965) 111–119.

[3] Cullender, M.H.: "The Isochronal Performance Method of Determining the Flow Characteristics of Gas Wells," *Trans.*, AIME (1955) 204, 137–142.

[4] Katz, D.L., Cornell, D., Kobayashi, R., Poettmann, F.H., Vary, J.A., Elenbaas, J.R., and Weinaug, C.F.: *Handbook of Natural Gas Engineering*, McGraw-Hill Book Co., Inc., New York (1959).

Chapter 2

Additional Testing Options

A closer look at the interpretation of traditional gaswell tests leads us to modify some of the testing procedures in order to gain more information about the formation near the wellbore. Also, the AOF plots, discussed in the preceding section, may be misleading because the test does not separate the effects of non-Darcy flow from those due to limited entry and wellbore damage. In this section, we shall introduce a more comprehensive treatment of the gasflow system by including factors for the near wellbore skin and turbulence flow effects.

2.1 Pseudopressure (Real Gas Potential) Treatment for Gas Flow

The accuracy of the inflow relationships that are obtained from gaswell testing techniques will depend on the validity of the assumptions and approximations made in the preceding sections. It can be shown that C and n, which we assumed to be constants, are in reality functions of such properties as viscosity, temperature, gas compressibility factor, reservoir permeability, net pay thickness, wellbore damage, wellbore radius, drainage radius, skin value, turbulence, and tortuosity of pores. If these factors do not change appreciably during the producing period of the field, the same inflow performance plot may be used for the life of the well. In practice, the gas properties usually change and so do n and C. Furthermore, we have to remember that we used a rather weak approximation in evaluating the integrals leading to equation (1.9) where viscosity and compressibility factors (which are strongly dependent upon pressure) were evaluated at an average pressure and, thereafter, were treated as constants.

In order to incorporate the changes that occur in gas properties with pressure, we introduce the pseudopressure or real gas potential concept as originally defined by Al-Hussainy and Ramey.[1]

$$\psi(p) = 2 \int_{p_b}^{p} \left[\frac{p}{\mu_g(p)z(p)} \right] dp \quad (2.1)$$

Where p_b is an arbitrary base pressure, and the p's in parentheses indicate that ψ, μ_g, and z are all functions of pressure.

By differentiating both sides of equation (2.1) and, using the fundamental theorem of calculus*, we can obtain

$$d\psi(p) = 2 \frac{p}{\mu_g(p)z(p)} dp \quad (2.2)$$

or

$$dp = \frac{\mu_g(p)z(p)}{2p} d\psi(p) \quad (2.3)$$

Integration is the reverse process of differentiation. Using equation (2.3), one can rewrite equation (1.4) as,

$$\int_{r_w}^{r_e} \left[\frac{1}{r} - \frac{r}{r_e^2} \right] dr = 2\pi hk \int_{\psi_{wf}}^{\psi_e} \left[\frac{p}{\mu_g(p)z(p)} \frac{\mu_g(p)z(p)}{2p} \right] d\psi(p) \quad (2.4)$$

---

* $\frac{d}{dt} \int_{a}^{t} f(x)dx = f(t)$

Note that limits of the integral on the right side of the above equation are changed to pseudopotentials too. Equation (2.5) can be integrated directly without making any approximations. Thus,

$$q_{sc} = \frac{0.703 \times 10^{-6} \, kh \, (\psi_e - \psi_{wf})}{T_R ln \left(\frac{r_e}{r_w} \right) - 0.75} \qquad (2.5)$$

Letting

$$C' = \frac{0.703 \times 10^{-6} kh}{T_R ln \left[\left(\frac{r_e}{r_w} \right) - 0.75 \right]} \qquad (2.6)$$

the corresponding form of equation (1.11) will be

$$q_{sc} = C' \, (\psi_e - \psi_{wf})^n \qquad (2.7)$$

The above equation (2.7) is very similar to equation (1.11) except that the pressure squared variables are replaced by the pseudopressure. It must be noted that the above equation was derived without making approximations for properties that are functions of pressure.

As the above development indicates, one needs a ψ versus p conversion table or plot (remember the limits of the integral have also been changed to ψ_e and ψ_{wf}). Once this table is obtained, any transformation between ψ and p can be easily made. For a given gas, the ψ versus p conversion table can be then used for the temperature for which it was developed. Since most gas reservoirs are isothermal and gas composition does not vary spatially in most cases, one ψ versus p conversion table will suffice for the entire reservoir.

It is important to establish conditions under which the pressure-squared approach may be used and those under which the most complicated pseudopressure approach should be used.

If a plot of $[\mu_g(p)\, z(p)]$ versus p is drawn for a natural gas at isothermal conditions, the graph should look like that shown in figure 2.1. The curve given in this figure shows that at low pressures, say up to 2000 psia, the product $[\mu_g(p)\, z(p)]$ is relatively constant, while at high pressures, above 5000 psia, it is essentially directly proportional to pressure. Keeping the above observation in mind, we can proceed as follows:

For a typical gas at low pressure when

$$\mu_g z = \text{constant} = \mu_{gi} z_i$$

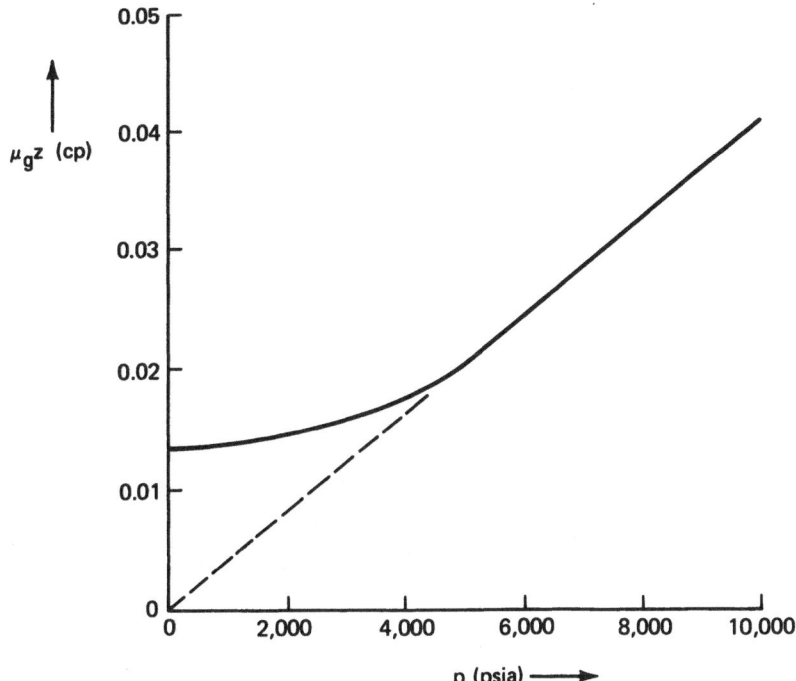

Figure 2.1
Plot of $\mu_g z$ Versus p for a Natural Gas.

$$\psi = \frac{2}{\mu_{gi}z_i} \int_{p_b}^{p} p\,dp$$

and if our base pressure is chosen as zero

$$\psi = \frac{1}{\mu_{gi}z_i} p^2 \qquad (2.8)$$

From equation (2.8) we can conclude that the flow equation in pressure-squared is valid for gas at pressures less than about 2000 psia.

On the other hand, at higher pressures, figure 2.1 shows that the slope of the curve is almost constant, that is

$$\frac{p}{\mu_g z} = \frac{p_i}{\mu_{gi}z_i}$$

so that

$$\psi = \frac{2p_i}{\mu_{gi}z_i} \int_{p_b}^{p} dp$$

and if our base pressure, p_b, is chosen as zero

$$\psi = \frac{2p_i}{\mu_{gi}z_i} p \qquad (2.9)$$

Equation (2.9) indicates the possibility of using a pressure (not pressure-squared) relationship at high pressures. This is to be expected because the behavior of gases at relatively high pressures is very similar to the behavior of liquids. However, extreme caution has to be taken in making generalizations of this sort.

One final note on the construction of the ψ versus p conversion table should be made. Using the experimental values of μ_g and z, the values of $2p/\mu_g z$ are calculated for several values of p. Then $2p/\mu_g z$ versus p is plotted and the area under the curve is calculated either numerically or graphically, where the area under the curve from $p = 0$ to any pressure, p, represents the value of ψ corresponding to p. The following example will illustrate the procedure.

Example 2.1
Construction of a ψ versus p
Conversion Table

The following data are available on a reservoir gas sample. Obtain a ψ versus p conversion chart. Employ the graphic integration technique.

p (psia)	μ_g(cp)	z
400	0.01286	0.937
800	0.01390	0.882
1200	0.01530	0.832
1600	0.01680	0.794
2000	0.01840	0.770
2400	0.02010	0.763
2800	0.02170	0.775
3200	0.02340	0.797
3600	0.02500	0.827
4000	0.02660	0.860
4400	0.02831	0.896

Solution
The pseudopressure plot is obtained by first preparing the following table:

p (psia)	μ_g(cp)	z	$\dfrac{2p}{\mu_g z}\left(\dfrac{\text{psia}}{\text{cp}}\right)$
400	0.01286	0.937	66391
800	0.01390	0.882	130508
1200	0.01530	0.832	188537
1600	0.01680	0.794	239894
2000	0.01840	0.770	282326
2400	0.02010	0.763	312983
2800	0.02170	0.775	332986
3200	0.02340	0.797	343167
3600	0.02500	0.827	348247
4000	0.02660	0.860	349711
4400	0.02831	0.896	346924

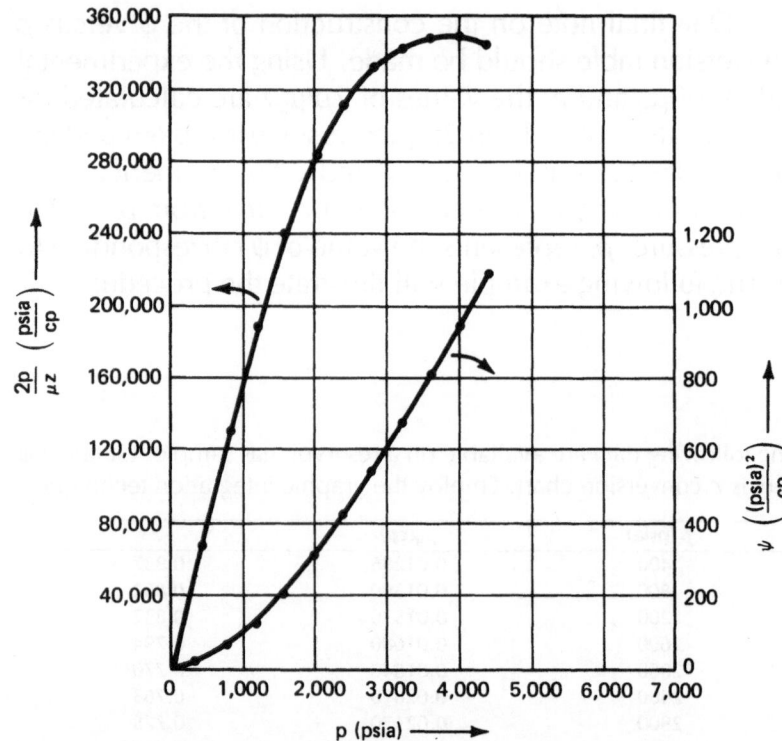

Example 2.1
Plot of data for Example 2.1.

Now, we construct a plot of $\dfrac{2p}{\mu_g z}$ versus p and calculate the area under the curve to each p value. Next we calculate the area under the curve for different values of p. This is shown in the accompanying figure. Areas calculated in this way will give the corresponding ψ values. These ψ values are tabulated below (ψ versus p is also plotted in the figure).

p (psia)	$\psi\left(\dfrac{\text{psia}^2}{\text{cp}}\right)$
400	13.2×10^6
800	52.0×10^6
1200	113.1×10^6
1600	198.0×10^6
2000	304.0×10^6
2400	422.0×10^6
2800	542.4×10^6
3200	678.0×10^6
3600	816.0×10^6
4000	950.0×10^6
4400	1089.0×10^6

2.2
The Radial Diffusivity Equation For Gas Flow

Our aim in this section is to develop a partial differential equation describing the transient flow behavior of a single-phase gas flowing through a porous medium. This equation is known as the diffusivity equation since it has the same form as that of the equation used to describe unsteady heat flow and unsteady mass diffusion. The following assumptions will be applied throughout the theoretical development:

(1) Darcy's law applies.
(2) Single-phase real gas flow exists.
(3) Gravitational effects are negligible.
(4) The formation is homogeneous, isotropic, and horizontal, and the permeability and porosity are constant.

We start with the continuity equation in radial form,

$$- \frac{1}{r} \frac{\partial}{\partial r} (r \rho_g v_r) = \frac{\partial}{\partial t} (\phi \rho_g) \qquad (2.10)$$

where equation (2.10) is developed in Appendix A. To derive the partial differential equation for fluid flow in a porous medium, we combine equation (2.10) with Darcy's law by simply defining the velocity in the r direction as

$$v_r = - \frac{k_r}{\mu_g} \frac{\partial p}{\partial r} \qquad (2.11)$$

Note that k_r is the permeability to flow in the r − direction. Substitution of the above equation into equation (2.10) yields

$$\frac{1}{r} \frac{\partial}{\partial r} \left[\frac{r \rho_g k_r}{\mu_g} \frac{\partial p}{\partial r} \right] = \frac{\partial}{\partial t} (\phi \rho_g) \quad (2.12)$$

The density, ρ_g, of the real gas will be obtained by the gas law as follows:

$$pV = \frac{m}{M} z RT \quad (2.13)$$

where

    V = volume occupied by the mass "m" of gas
    M = molecular weight
    R = gas constant
    T = absolute temperature
    p = pressure

Since $\rho = \frac{m}{V}$, in this case the density of a real gas is given by

$$\rho_g = \frac{M}{RT} \frac{p}{z} \quad (2.14)$$

Substituting equation (2.14) into equation (2.12) and treating permeability, k_r, and porosity, ϕ, as constants, one obtains

$$\frac{1}{r} \frac{\partial}{\partial r} \left[\frac{M}{RT} \frac{pr}{\mu z} \frac{\partial p}{\partial r} \right] = \frac{\phi}{k} \frac{M}{RT} \frac{\partial}{\partial t} \left[\frac{p}{z} \right]$$

or

$$\frac{1}{r} \frac{\partial}{\partial r} \left[\frac{p}{\mu_g z} r \frac{\partial p}{\partial r} \right] = \frac{\phi}{k} \frac{\partial}{\partial t} \left[\frac{p}{z} \right] \quad (2.15)$$

At this point we again calculate $\bar{\mu}_g$ and $\bar{z}$ at an average pressure $\bar{p}$ where

$$\bar{p} = (\bar{p}_R^2 + p_{wf}^2)^{1/2}$$

Then, equation (2.15) takes the following form

$$\frac{1}{r} \frac{\partial}{\partial r} \left(r \; \frac{\partial (p)^2}{\partial r} \right) = \frac{\phi \bar{\mu}_g}{kp} \frac{\partial (p)^2}{\partial t} \qquad (2.16)$$

Expanding the left-hand side of the above expression we obtain,

$$\frac{\partial^2 p^2}{\partial r^2} + \frac{1}{r} \frac{\partial p^2}{\partial r} = \frac{\phi \bar{\mu}_g}{kp} \frac{\partial p^2}{\partial t} \qquad (2.17)$$

Equation (2.17) is the radial form of the diffusivity equation that describes the isothermal, single-phase real gas flow in a homogeneous, isotropic porous medium.

2.3 Analytical Solution of the Diffusivity Equation

The solution of the diffusivity equation derived in the previous section should describe the pressure at any point in the radial flow system as a function of time. This equation has no known analytical solution because it is nonlinear. However, an approximate solution is derived in Appendix B for certain boundary and initial conditions. The approximate solution developed there is for a single well located at the center of a circular reservoir of infinite radial extent. The solution for the pressure at the wellbore, p_w, written in field units, without skin factor or turbulence, is given as follows:

$$p_R^2 - p_{wf}^2 = \frac{1637 \, \bar{\mu}_g \bar{z} \, T_R \, q_{sc}}{kh} \left[\log t + \log \left(\frac{k}{\phi \bar{\mu}_g c r_w^2} \right) - 3.23 \right] \qquad (2.18)$$

where:

p_R = initial average reservoir pressure, psia

p_{wf} = pressure at the wellbore at time t, psia

$$\bar{p} = \frac{P_R - P_{wf}}{2}$$

$\bar{\mu}_g$ = gas viscosity at average pressure $\bar{p}$, cp

$\bar{z}$ = gas compressibility factor at average pressure $\bar{p}$

T_R = reservoir temperature, °R

k = permeability, md

h = formation thickness, ft

c = gas compressibility, psi^{-1} or $1/p$

ϕ = porosity, fraction

r_w = well radius, ft

t = time, hrs

q_{sc} = production rate, MSCF/D

2.4 Flow Regimes

Several testing procedures and analysis techniques have been developed to examine the pressure distribution with time in the flow system, and, more important, to gather valuable information on near wellbore and/or reservoir characteristics. Analysis techniques are based on the solutions for radial flow of a single-phase fluid similar to the one presented in the preceding section.

One such testing procedure, developed in connection with the utilization of the analytical solution of the diffusivity equation, is to shut in a well until its pressure has stabilized, open it to flow at constant rate, and observe its pressure behavior at the wellbore. This test is known as a *drawdown test*. A schematic plot of the wellbore pressure for a well producing under such conditions is given in figure

2.2. As shown in figure 2.2, and providing the reservoir is finite in radial extent, we may recognize three states of flow.

(1) Unsteady state or transient.
(2) Transitional or late transient.
(3) Pseudosteady state or quasisteady state.

The approximate end of the unsteady state flow period is given by:

$$t \cong 360 \; \frac{\phi \mu_g c r_e^2}{k} \qquad (2.19)$$

and the start of pseudosteady state flow period by:

$$t \cong \frac{1400 \phi \mu_g c r_e^2}{k} \qquad (2.20)$$

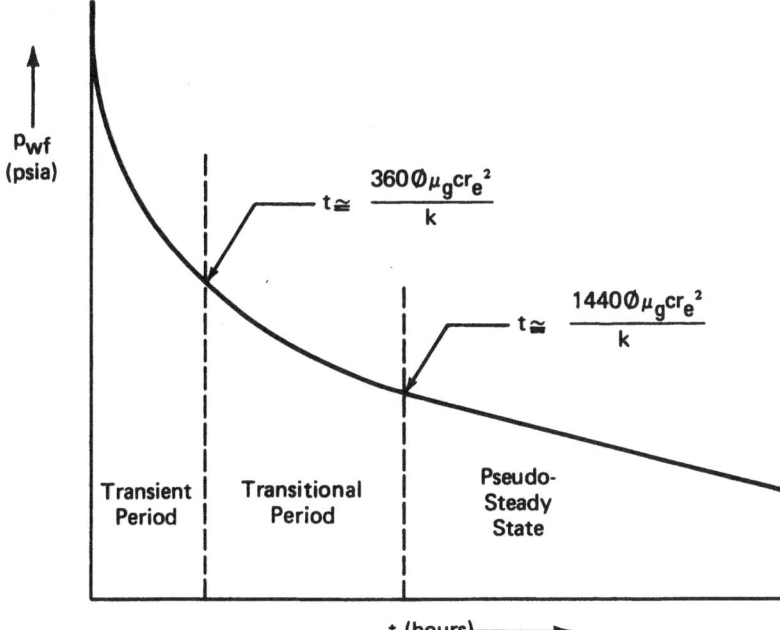

Figure 2.2
Observed States of Flow during a Drawdown Test (Finite System).

These approximations are discussed by Odeh and Nabor.[2]

During pseudosteady state flow, the change of pressure with time is constant. This means that a plot of p_{wf} versus time will become a straight line on linear coordinate paper during pseudosteady state flow.

If the reservoir is large (infinite in radial extent) unsteady state flow will persist and both transitional and pseudosteady state flow will not be observed.

An analysis technique applicable to the unsteady state flow period will be presented later.

2.5
Skin Factor

Equation (2.18), the solution to the radial diffusivity equation, gives the pressure distribution in the reservoir for the basic flow system under ideal conditions. It is well known that the properties of the formation near the wellbore are usually altered during drilling, completion, and stimulation procedures. Invasion by drilling fluids, the presence of mud cake and cement, partial well penetration, and limited entry perforations are some of the factors that cause damage to the formation; and, hence, an additional localized pressure drop during flow. On the other hand, well stimulation techniques, such as acidizing and fracturing, will normally enhance the properties of the formation and increase the permeability around the wellbore, so that a decrease in pressure drop over that otherwise expected for a given flow rate is observed. Therefore, with the basic flow system and with our basic solution, we should incorporate the additional pressure effects caused by near-wellbore differences in formation properties. The zone of altered permeability is referred to as a skin and the resulting effect as a skin effect. Figure 2.3 schematically shows the additional pressure

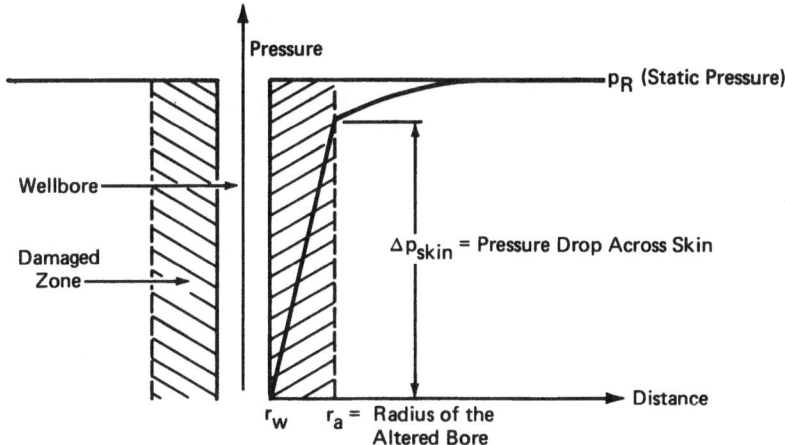

Figure 2.3
Schematic Representation of Near Wellbore Skin Effect.

drop caused by a damaged zone in the vicinity of the wellbore.

Figure 2.4 compares the differences in pressure distribution in a formation of constant permeability, with that in a formation with an improved zone near the wellbore and that in a formation with a damaged zone near the wellbore.

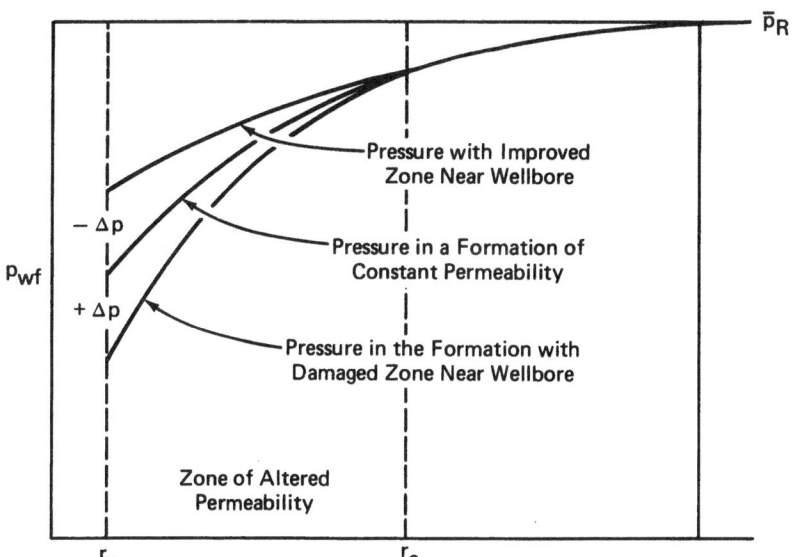

Figure 2.4
Schematic Representation of Positive and Negative Skin Effects.

Hawkins[3] showed that the skin can be expressed by:

$$S = (k/k_a - 1) \, \ln (r_a/r_w) \qquad (2.21)$$

where k_a is the permeability in the zone of altered permeability. When a damaged zone near the wellbore exists, k_a is less than k, hence S is a positive number (note that $\ln(r_a/r_w)$ is always positive). If there is an improved zone near the wellbore, k_a is greater then k causing S to be a negative number. In other words, a positive value of S indicates a damaged well, and a negative value indicates an improved well.

We can incorporate the effect of skin into our basic solution, equation (2.18).

$$q_{sc} = \frac{0.703 \times 10^{-3} \, kh \, (p_a{}^2 - p_{wf}{}^2)}{\bar{\mu}_g T_r \, \bar{z} \, \ln \left(\dfrac{r_a}{r_w} \right)} \qquad (2.22)$$

Where q_{sc} is in MSCF/D, k is the original permeability, and p_a in the pressure at r_a. Rearranging equation (2.22) for $(p_a{}^2 - p_{wf}{}^2)$ one obtains,

$$(p_a{}^2 - p_{wf}{}^2)_1 = (\Delta p^2)_1 = \frac{q_{sc}\bar{\mu}_g T_R \bar{z} \, \ln \left(\dfrac{r_a}{r_w} \right)}{0.703 \times 10^{-3} \, kh} \qquad (2.23)$$

Similarly the pressure drop within the zone of altered permeability is

$$(\Delta p^2)_2 = \frac{q_{sc}\bar{\mu}_g T_R \bar{z} \, \ln \left(\dfrac{r_a}{r_w} \right)}{0.703 \times 10^{-3} \, k_a \, h} \qquad (2.24)$$

Thus $(\Delta p^2)_{skin} = (\Delta p^2)_2 - (\Delta p^2)_1$

or $(\Delta p^2)_{skin} = \dfrac{q_{sc}\bar{\mu}_q T_R \bar{z}}{0.703 \times 10^{-3} h} \ln\left(\dfrac{r_a}{r_w}\right)\left[\dfrac{1}{k_a} - \dfrac{1}{k}\right]$

or $(\Delta p^2)_{skin} = \dfrac{q_{sc}\bar{\mu}_g T_R \bar{z}}{0.703 \times 10^{-3} kh} \ln\left(\dfrac{r_a}{r_w}\right)\left[\dfrac{k - k_a}{k_a}\right]$ (2.25)

Incorporating Hawkins' definition for skin, where S is the skin factor, equation (2.25) can be written as:

$$(\Delta p^2)_{skin} = \dfrac{q_{sc}\bar{\mu}_g T_R \bar{z}}{0.703 \times 10^{-3} kh}\, (S) \quad (2.26)$$

Now, we can add this supplemental pressure drop to equation (2.18) and obtain:

$$p_R^2 - p_{wf}^2 = \dfrac{1637\bar{\mu}_g \bar{z} T_R q_{sc}}{kh}$$

$$\left[\log t + \log\left(\dfrac{k}{\phi\bar{\mu}_g c r_w^2}\right) - 3.23 + 0.87\, S\right] \quad (2.27)$$

Remember that the above refers only to S, that is, the effect of the skin due to changes in permeability around the wellbore. The total skin, S', in reality, is the composite effect of limited entry to flow and formation damage or improvement near the wellbore. The engineer must discriminate between these two effects before it can be ascertained that a stimulation or perforation job will be helpful. Estimating the skin due to restricted entry to flow is discussed in Appendix C.

2.6 Turbulent Flow Factor

All of the mathematical formulations presented so far are based on the assumption that laminar flow conditions are observed during flow. During radial flow, the flow velocity increases as the wellbore is approached, because the cross-sectional area perpendicular to the flow direction becomes smaller. This increase in velocity will favor the development of turbulent flow around the wellbore. If turbulent flow does exist, and it is most likely to occur with gases, an additional pressure drop similar to that caused by a skin effect, but one that is rate sensitive, will be present.

In a manner similar to equation (2.26), one can write the following equation to describe the additional pressure drop caused by turbulent flow where D is the turbulent flow factor:

$$(\triangle p^2)_{\text{turbulent}} = \frac{q_{sc}\bar{\mu}_q T_R \bar{z}}{0.703 \times 10^{-3} \, kh}(Dq_{sc}) \qquad (2.28)$$

Where D is the turbulent flow factor.

Incorporating the above additional pressure drop into equation (2.27) yields:

$$p_R^2 - p_{wf}^2 = \frac{1637\bar{\mu}_g \bar{z} T_R q_{sc}}{kh}\left[\log t + \log\left(\frac{k}{\phi\mu_g cr_w^2}\right) - 3.23 + 0.87S + 0.87Dq_{sc}\right] \qquad (2.29)$$

Equation (2.29), then, includes the pressure effects of both physical skin and turbulent flow (non-Darcy flow skin). We recognize that both of these skin factors occur in the vicinity of the wellbore and are detected as a single effect. Accordingly, we define the composite skin factor, S', as:

$$S' = S + Dq_{sc} \qquad (2.30)$$

and rewrite equation (2.29) as follows:

$$p_R^2 - p_{wf}^2 = \frac{1637\bar{\mu}_g \bar{z} T_R q_{sc}}{kh}\left[\log t + \log\left(\frac{k}{\phi\mu_g c r_w^2}\right) - 3.23 + 0.87S'\right] \qquad (2.31)$$

Equation (2.31) is valid for infinite systems. It is also valid for finite systems until the effect of an outer boundary is felt—until the end of transient flow occurs.

2.7
Drawdown Or Falloff Test

In looking at our basic equation, equation (2.31), we realize that certain terms are known from basic data, specifically, μ_g, $\bar{z}$, T_R, h, ϕ, c, r_w; and other terms are measured during a well test, specifically, $\bar{p}_R$, p_{wf}, q_{sc}, t. This leaves us with three unknowns: permeability, k or, kh, the product of permeability and thickness, if thickness is not known; physical skin factor, S or S'; and the non-Darcy flow coefficient, D. Testing procedures that have been developed to determine these unknowns are discussed in the remaining sections of this chapter.

A pressure drawdown test is conducted by placing a well on production at a constant flow rate and observing the pressure in the wellbore over time. At the start of the test, the bottomhole pressure should be equal to the static reservoir pressure. This is achieved by shutting in the well and allowing it to reach a stabilized pressure. This pressure is assumed to be equal to the static reservoir pressure. The well is then opened to production and allowed to flow at a constant rate. The flowing bottomhole pressures are measured and recorded as a function of time during this flow period.

Falloff tests are carried out on injection wells. The falloff in bottomhole pressure is recorded after injection is terminated.

If we take a closer look at our generalized solution, equation (2.31), we note that if the flow rate, q_{sc}, is constant during a drawdown, a plot of p_{wf}^2 versus log t will give a straight line (figure 2.5), which has a slope:

$$m = -\frac{1637\bar{\mu}_g\bar{z}T_R q_{sc}}{kh} \qquad (2.32)$$

and since q_{sc}, T_R, $\bar{\mu}_g$, $\bar{z}$ and, quite likely, h, are known, it is a simple calculation to find k. The slope, m, in equation (2.32) is negative. This occurs because p_{wf}^2 decreases as t increases. The slope m has units of psia²/cycle, and is measured as pressure drop squared per cycle. A rearrangement of equation (2.32) gives the solution for permeability as follows:

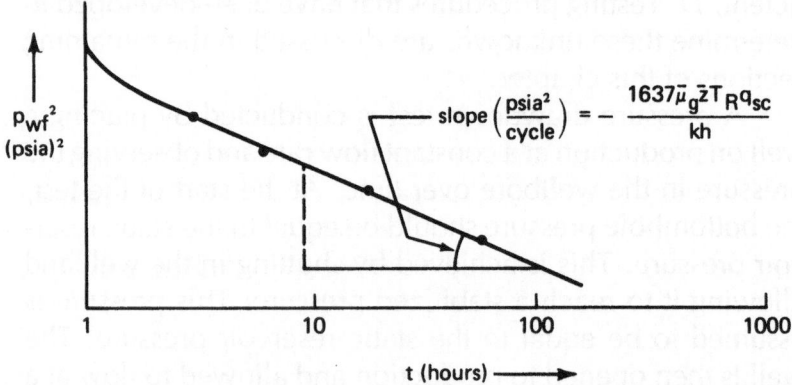

Figure 2.5
Typical Plot of Pressure Versus Time during a Drawdown Test.

$$k = -\frac{1637\bar{\mu}_g \bar{z} T_R\, q_{sc}}{mh} \qquad (2.33)$$

Since m is negative, the value of permeability will be positive.

2.8
Buildup Test

A pressure buildup test is one in which a well is produced at a constant rate for a period of time t, and then shut in. The pressure increase in the wellbore after shut-in is measured as a function of time. This test is similar to the drawdown test in that it allows for the calculation of formation permeability. In fact, it is usually preferable because it is difficult to achieve a constant rate of flow during a drawdown test. In the following pages, the fundamental relationships and applications of the buildup test are developed.

Conceptually, a buildup test may be considered as the result of two superposed effects: a pressure drawdown caused by the constant production rate q_{sc} assumed to occur for a time, $t_p + \Delta t$; but, at the time of shut in, t_p, a hypothetical second drawdown at a rate of $-q_{sc}$ is initiated and continues for a time, Δt. The net effect of the hypothetical negative flow rate (which corresponds to injection in our notation) imposed on the positive flow rate is to simulate a flow rate of zero, which is the shut-in condition. Thus, at any shut-in time, Δt, the pressure behavior at the well will be the superposed effects of flow rate q_{sc} for a time $(t_p + \Delta t)$ and flow rate $-q_{sc}$ for a time, Δt. The application of the principle of superposition is shown in figure 2.6. For infinite systems the application of the superposition principle gives the following expression:

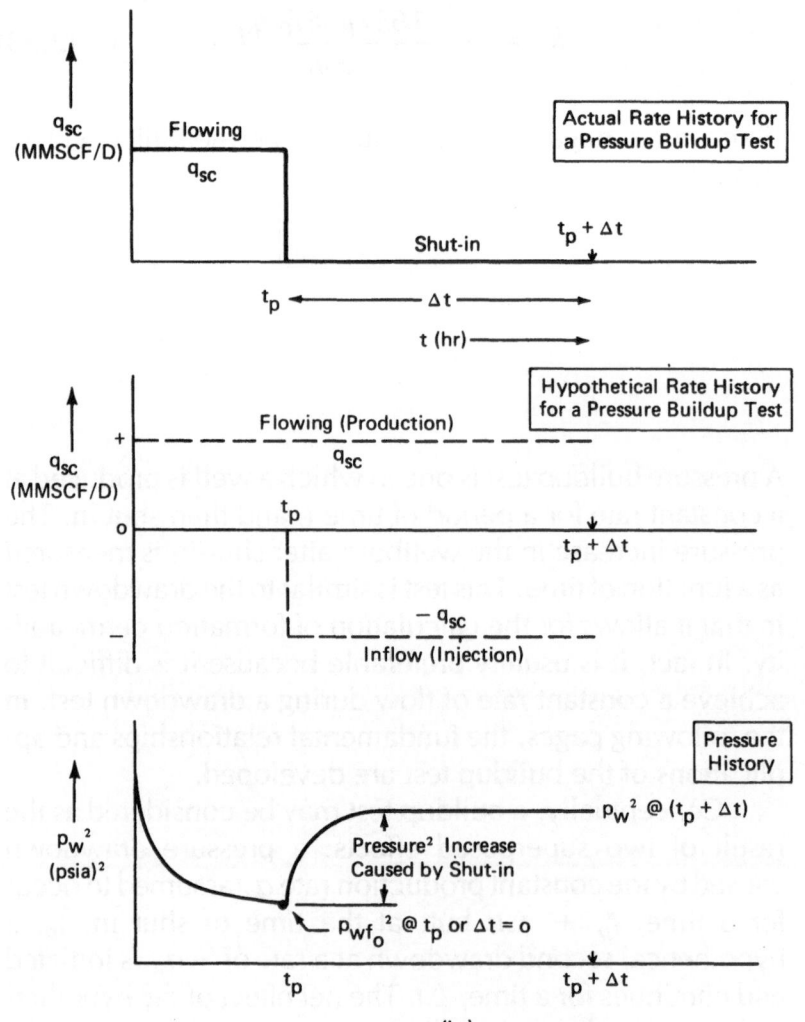

Figure 2.6
Flow Rate and Pressure History of a Pressure Buildup Test.

$$p_R^2 - p_w^2 = \left\{ \frac{1637\bar{\mu}_g \bar{z} T_R q_{sc}}{kh} \left[\log(t_p + \Delta t) + \log\left(\frac{k}{\phi \mu_g c r_w^2}\right) - 3.23 \right] \right\}$$

$$+ \left\{ -\frac{1637\bar{\mu}_g \bar{z} T_R q_{sc}}{kh} \left[\log \Delta t + \log\left(\frac{k}{\phi \mu_g c r_w^2}\right) - 3.23 \right] \right\} \quad (2.34)$$

44 Gaswell Testing

It should be recognized that the first curly brackets represent the pressure-squared drop caused by producing the well at a rate q_{sc} for a period of $t_p + \Delta t$, and the second curly brackets represent the pressure-squared buildup due to the hypothetical injection at a rate $-q_{sc}$ for a period of Δt. Also, note that the composite skins should not be superposed in time since they are only a function of the existing flow rates.

Equation (2.34) can be rewritten as:

$$p_R^2 - p_w^2 = \frac{1637 \bar{\mu}_g \bar{z} T_R q_{sc}}{kh} \left[\log \left(\frac{t_p + \Delta t}{\Delta t} \right) \right] \quad (2.35)$$

The producing time, t_p, is usually approximated by taking the rate q_{sc} as the last rate before closing in the well and dividing the cumulative well production since completion, or since a prolonged period of shut-in, by this rate,

$$t_p = \frac{\text{cumulative production}}{\text{production rate before shut-in}} \quad (2.36)$$

Since t is in hours in equation (2.35), the units of equation (2.36) should be chosen accordingly. Rearranging equation (2.35) one obtains:

$$p_w^2 = p_R^2 - \frac{1637 \bar{\mu}_g \bar{z} T_R q_{sc}}{kh} [\log \left(\frac{t_p + \Delta t}{\Delta t} \right)] \quad (2.37)$$

Now it is apparent that a plot of p_w^2 versus $\log \frac{t_p + \Delta t}{\Delta t}$ is a straight line with a slope of:

$$m = - \frac{1637 \bar{\mu}_g \bar{z} T_R q_{sc}}{kh} \quad (2.38)$$

Where $m = $ slope, $(\text{psia})^2/\text{cycle}$.

A plot of p_w^2 (or p) versus $\log \dfrac{t_p + \Delta t}{\Delta t}$ is given in figure 2.7. As the figure indicates, the slope of the straight line is negative; therefore, when equation (2.38) is rearranged to read:

$$k = - \frac{1637 \bar{\mu}_g \bar{z} T_R q_{sc}}{mh} \qquad (2.39)$$

it will give a positive value for permeability.

For an infinite shut-in time, Δt_∞, the value of $\dfrac{t_p + \Delta t_\infty}{\Delta t}$ is equal to one. Practically, this corresponds to zero pressure drop as Δt goes to infinity. This simply means that by extrapolating the straight line of the buildup data to $\dfrac{t_p + \Delta t}{\Delta t} = 1$, we should get the static reservoir pressure for an infinite reservoir. This is also shown in figure 2.7. One last but very important note is that the slope must

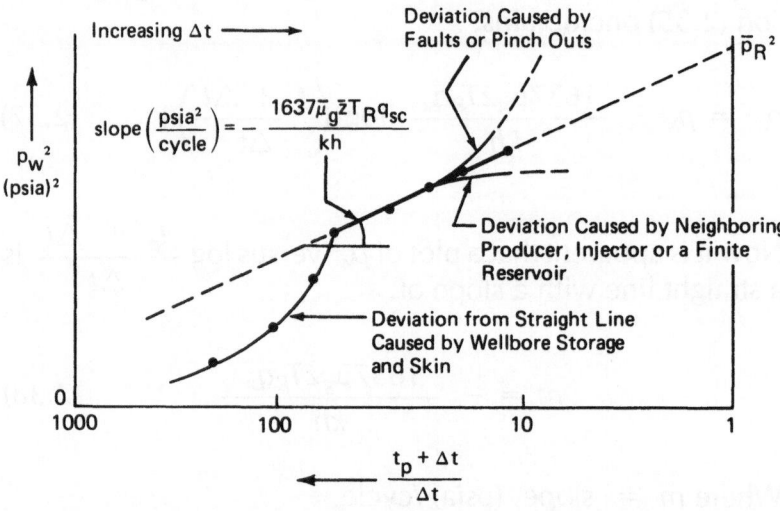

Figure 2.7
Semi-log Plot of Buildup Data.

46   Gaswell Testing

always be measured on the straight line portion of the curves, not during the early shut-in periods when the straight line does not exist.

A change in slope along the end portion of the straight line may indicate reservoir discontinuities, such as faults or pinchouts, or may be caused by neighboring producers, injectors, or a finite reservoir.

2.9 Determination of S, S′ and D from Buildup and Drawdown Tests

If we define $p_{wf_o}^2$ as the pressure-squared just prior to shut-in, that is, at $\Delta t = 0$ (see figure 2.6); then we can write equation (2.31) as:

$$p_R^2 - p_{wf_o}^2 = \frac{1637\bar{\mu}_g \bar{z} T_R q_{sc}}{kh} \left[\log t_p + \log \left(\frac{k}{\phi \mu_g c r_w^2} \right) - 3.23 + 0.87S' \right] \quad (2.40)$$

Then, rewriting equation (2.35) from equation (2.40), one obtains:

$$p_w^2 - p_{wf_o}^2 = \frac{1637\mu_g \bar{z} T_R q_{sc}}{kh} \left[\log \left(\frac{t_p}{t_p + \Delta t} \right) + \log \Delta t + \log \right.$$

$$\left. \left(\frac{k}{\phi \mu_g c r_w^2} \right) - 3.23 + 0.87S' \right] \quad (2.41)$$

Defining $p_w^2\ (\triangle t = 1)$ as the pressure-squared at $\triangle t = 1$ hour, and making the approximation of $\dfrac{t_p}{t_p + 1} \cong 1$, then equation (2.41) may be simplified to give:

$$S' = 1.15 \left[\frac{p_w^2\ (\triangle t = 1) - p_{wf_o}^2}{-m} - \log\left(\frac{k}{\phi\mu_g c r_w^2}\right) + 3.23 \right] \qquad (2.42)$$

Earlier, we mentioned that as a result of using the superposition principle, the composite skin factors, S', does not appear in equation (2.35). Although S' does not appear in the general pressure buildup equation, it still affects the shape of the early pressure buildup data. In fact, an early-time deviation from a straight line will be observed with most pressure buildup data, see figure 2.7. We must keep in mind that the slope of the buildup curve and the value of $p_w^2\ (\triangle t = 1)$ of equation (2.42) should be obtained from the straight-line portion (extrapolate if necessary) of the pressure buildup plot.

Because S' has two unknowns, S and D, we must have two equations for a unique solution. It is virtually impossible to separate S' into its components S and Dq_{sc} from a single buildup test. However, a drawdown-buildup test at one rate, q_{sc_1}, followed by a second drawdown-buildup test at a second rate q_{sc_2} (q_{sc_2} must be substantially different from q_{sc_1}). It should be noted, however, that the well must be shut in before the second test and allowed to reach original pressure conditions. Figure 2.8 schematically shows the test cycle required.

The first drawdown-buildup cycle will give:

$$S'_1 = S + Dq_{sc_1} \qquad (2.43)$$

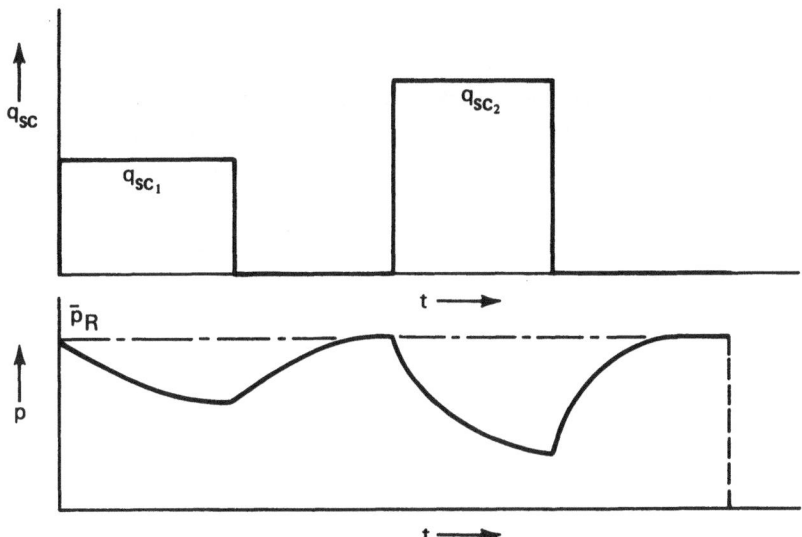

Figure 2.8
Two Cycle Drawdown—Buildup Test.

And the second drawdown-buildup cycle will give:

$$S'_2 = S + Dq_{sc_2} \qquad (2.44)$$

Solving equations (2.43) and (2.44) simultaneously for D and S one obtains:

$$D = \frac{S'_1 - S'_2}{q_{sc_1} - q_{sc_2}} \qquad (2.45)$$

and

$$S = \frac{q_{sc_2} S' - q_{sc_1} S'_2}{q_{sc_2} - q_{sc_1}} \qquad (2.46)$$

It should be recognized that the two drawdown-buildup test sequences are equivalent to the first two flow rates in an isochronal test, except that in the drawdown-buildup cycle, we must measure pressure continuously.

Although the composite skin factor could have been obtained using two drawdown-buildup tests, it is generally preferred within the industry that buildup tests of at least three cycles be used to obtain three data points. This testing procedure is equivalent to a three-cycle isochronal test. Then, S' versus q_{sc} is plotted on linear coordinates and the best straight line is drawn through the points, see figure 2.9.

The slope of this line is equal to D, and the intercept at $q_{sc} = 0$ is equal to S.

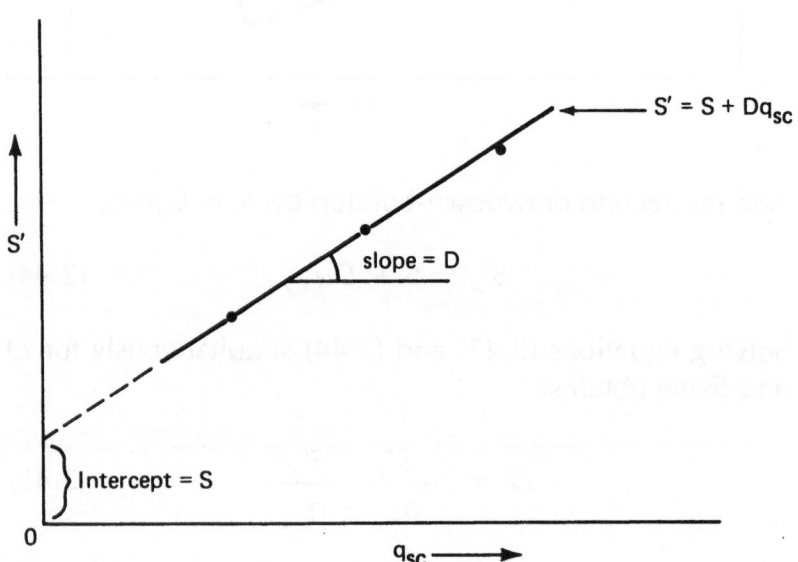

Figure 2.9
Graphic Determination of S and D.

Example 2.2
Calculation of the Composite Skin Factors

From the analysis of a three-cycle buildup test, the following S' values are obtained for the respective values of q_{sc}. Find the physical skin, S, and turbulent flow factor, D.

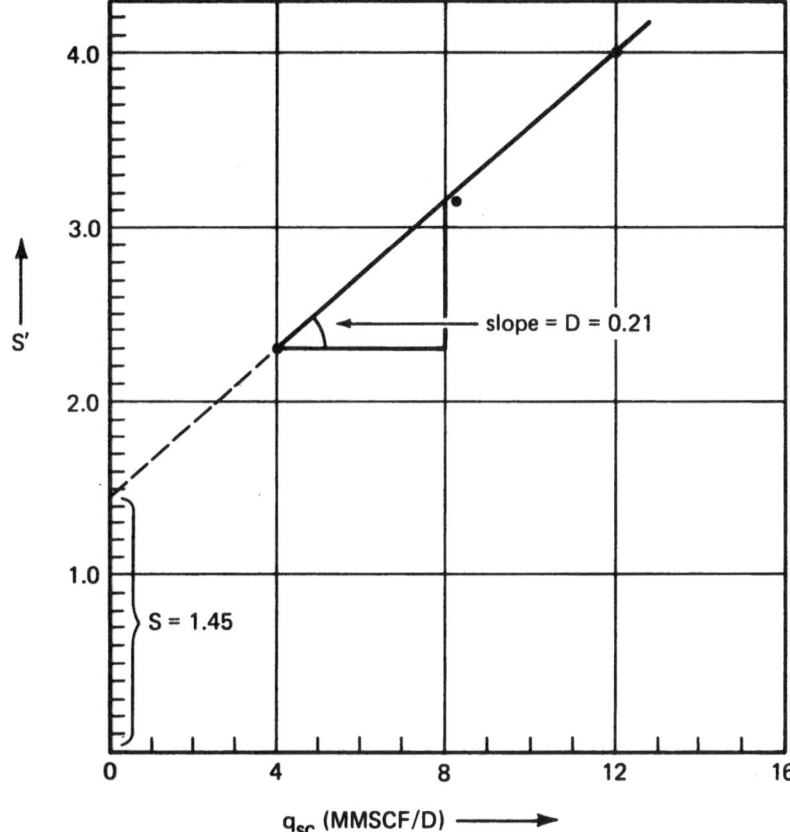

Example 2.2
Graphic Calculation of S and D.

S'	q_{sc} (MMSCF/D)
2.30	4.0
3.15	8.2
4.0	12.0

Solution
S' versus q_{sc} is plotted on linear coordinate paper and the best straight line is drawn through the points. The equation of the line is:

$$S' = S + Dq_{sc}$$

Therefore, the slope of the line will give the value of D and the intercept on the S' axis will give the value of S. Thus, as shown in the following plot:

$$S = 1.45$$

$$D = 0.21 \text{ D/MMSCF}$$

References

[1] Al-Hussainy, R. and Ramey, H.J., Jr.: "Application of Real Gas Flow Theory to Well Testing and Deliverability Forecasting," *J. Pet. Tech.* (May 1966) 637–642.

[2] Odeh, A.S. and Nabor, G.W.: "The Effect of Production History on Determination of Formation Characteristics from Flow Tests," *J. Pet. Tech.* (October 1966) 1343–1350.

[3] Hawkins, M.F., Jr.: "A Note on the Skin Effect," *Trans.*, AIME (1956) 207, 356–357.

Chapter 3

Additional Testing Considerations

Before proceeding to test-design principles, testing procedures, regulations, and reporting of a specific gaswell test, there are several further considerations to be mentioned. Specifically, we shall discuss: the use of surface pressures in our relationships, time to stabilization, radius of investigation, choice of flow rates, duration of flow rates, wellbore storage effects, production of liquids, choice of measurement and well testing equipment, sampling, and safety in testing.

For further discussion of these topics see *Gas Well Testing: Theory and Practice.*[1]

3.1 Surface-Pressure Relationships

Deliverability relationships using wellhead rather than sandface pressures are useful because they are directly related to surface conditions, that is, to the gathering system backpressure. However, because the difference in pressure between the surface and sandface depends on downhole equipment and flow rates, a relationship founded on surface pressure will not be constant throughout the life of a well. It is recommended, especially at pressures above 2000 psia, that the fundamental relationship for the deliverability of a gas well be based on sandface pressures and that appropriate correlations, usually available in existing computer programs, be used to convert sandface pressures to surface pressures if surface conditions are needed.

It is critical that the engineer retain consistent pressure measurement location (surface or sandface) throughout the interpretive life of the well. Many times it is important to convert surface pressures to sandface (bottomhole) and vice versa. In one-phase, liquid systems, this conversion is quite simple. For example, the sandface pressure for a liquid-saturated wellbore of reasonably constant density may be calculated using the following relationship:

$$p_{ws} = p_{ts} + \frac{\rho}{144} d \qquad (3.1)$$

Where:

p_{ws} = static sandface pressure, psia
p_{ts} = static wellhead pressure, psia
ρ = fluid density, lb/ft³
d = well depth, feet

This assumes that the liquid is incompressible or only slightly compressible.

Since gas density is a function of pressure, the above relationship does not hold for gases, and therefore an integral relationship taking into account the pressure-density relationship must be used. One presented by Cullender and Smith[2] is appropriate. It is written:

$$\frac{1000\, Gd}{53.34} = \int_{p_{tf}}^{p_{wf}} \frac{\left(\dfrac{p}{Tz}\right) dp}{\left[F^2 + 0.001 \left(\dfrac{p}{Tz}\right)^2 \right]} \qquad (3.2)$$

While this integral is difficult to solve analytically, by breaking it into parts and using the trapezoidal rule, an approximation solution is found. Then, equation (3.2) becomes:

$$\frac{1000\, Gd}{53.34} = \int_{p_{tf}}^{p_1} (\ ) + \int_{p_1}^{p_2} (\ ) + \int_{p_2}^{p_3} (\ ) + \ldots + \int_{p_{n-1}}^{p_n} + \int_{p_n}^{p_w} (\ ) \qquad (3.3)$$

Using only two parts, equation (3.3) becomes:

$$\frac{1000\,Gd}{53.34} = \int_{p_{tf}}^{p_{mf}} \frac{\left(\dfrac{p}{Tz}\right)}{\left[F^2 + 0.001\left(\dfrac{p}{Tz}\right)^2\right]}\,dp + \int_{p_{mf}}^{p_{wf}} \frac{\left(\dfrac{p}{Tz}\right)}{\left[F^2 + 0.001\left(\dfrac{p}{Tz}\right)^2\right]}\,dp \quad (3.4)$$

Now, using the trapezoidal rule, equation (3.4) becomes:

$$37.5\,Gd = (p_{mf} - p_{tf})(I_{mf} + I_{tf}) + (p_{wf} - p_{mf})(I_{wf} + I_{mf}) \quad (3.5)$$

Where:

$$I_n = \frac{\left(\dfrac{p}{Tz}\right)_n}{F^2 + 0.001\left(\dfrac{p}{Tz}\right)_n^2} \qquad n = mf,\ tf,\ wf \quad (3.6)$$

and

$$F = \frac{2.6665\,f\,q_{sc}^2}{d_i^5}$$

and f is the friction factor. Much experimental work, especially in horizontal pipes, has been carried out in order to determine the variables that affect f. For rough, long tubes (as in gas wells) the friction factor has been found to be a function of the Reynold's number and the relative roughness of the tube. The relative roughness has been defined as the ratio of the absolute roughness, δ (the distance between peaks and valleys in pipe-wall irregularities), to the internal diameter of the pipe, d_i. The absolute roughness of pipe will depend on whether it is clean or dirty and the material from which it is made (that is, plastic, commercial steel, etc.) for new pipe or tubing which is typically installed in gas wells. The absolute roughness is usually found in the range of 0.00060 to 0.00065. If the friction factor equation recommended by Nikuradse[3] for fully turbulent flow, assuming a friction factor of 0.00060, is used, the factor, F, becomes:

$$F = \frac{(0.10797)q_{sc}}{d_i^{2.612}} \, , \; d_i < 4.277 \text{ inches} \qquad (3.7)$$

$$F = \frac{(0.10337)q_{sc}}{d_i^{2.582}} \, , \; d_i > 4.277 \text{ inches} \qquad (3.8)$$

For other types of material or for dirty pipes the reader is referred to the work of Smith.[4]

Breaking equation (3.5) into two parts, we obtain for the upper half of the flow string:

$$37.5 \; G \frac{d}{2} \; = \; (p_{mf} - p_{tf})(I_{mf} + I_{tf}), \qquad (3.9)$$

and for the lower half of the flow string:

$$37.5 \; G \frac{d}{2} \; = \; (p_{wf} - p_{mf})(I_{wf} + I_{mf}) \qquad (3.10)$$

Applying Simpson's rule gives a more accurate value for the flowing bottomhole pressure

$$37.5 \; Gd \; = \; \frac{p_{wf} - p_{tf}}{3} \; (I_{tf} + 4I_{mf} + I_{wf}) \qquad (3.10a)$$

Where:

p = static or flowing pressure, psia
T = temperature, °R

z = gas compressibility factor, fraction
F = friction factor, zero if static
q_{sc} = flow rate, MMSCF/D
d_i = internal diameter of casing or tubing, inches
d = depth of well, ft
G = specific gravity of gas, relative to air, dimensionless

and the subscripts:

wf = flowing well
mf = middle of the flowing well
tf = top of flowing well

The following procedure is recommended by the Alberta Energy Resources Conservation Board[5] for the solution of equation (3.5):

(1) Calculate the left-hand side of equation (3.9) for the upper half of the flow string.
(2) Calculate F from equation (3.7) or (3.8), and then F^2.
(3) Calculate l_{tf} from equation (3.6) and wellhead conditions.
(4) Assume l_{mf} = l_{tf} for the conditions at the average well depth or at the midpoint of the flow string.
(5) Calculate p_{mf} from equation (3.9).
(6) Using the value of p_{mf} calculated in step 5, and the arithmetic average temperature, T_{mf}, determine the value of l_{mf} from equation (3.6).
(7) Recalculate p_{mf} from equation (3.9) and if this recalculated value is not within 1 psi of the p_{mf} calculated in step 5, repeat steps 6 and 7 until the above criterion is satisfied.
(8) Assume l_{wf} = l_{mf} for the conditions at the bottom of the flow string.

(9) Repeat steps 5 to 7, using equation (3.10) for the lower half of the flow string and obtain a value of the bottomhole pressure, p_{wf}.

(10) Apply Simpson's rule, equation (3.10a), to obtain a more accurate value of the flowing bottomhole pressure.

Example 3.1
Bottomhole Pressure
Calculation Using Cullender
and Smith Method

Calculate the flowing bottomhole pressure using the method of Cullender and Smith given the following data:

Gas gravity, G	= 0.74
Well depth, d	= 5000 ft
Flow rate, q_{sc}	= 2.500 MMSCF/D
Tubing inside diameter, d_1	= 2.441 in
Wellhead temperature, T_{tf}	= 515°R
Formation temperature, T_{wf}	= 540° R
Flowing wellhead pressure, p_{tf}	= 1000 psia
Pseudocritical temperature, T_c	= 408°R
Pseudocritical pressure, p_c	= 667 psia

Solution

$T_{mf} = (T_{tf} + T_{wf})/2 = (515 + 540)/2 = 528°R$

Wellhead,	$T_r = T_{tf}/T_c$	=	515/408	= 1.262
Midpoint,	$T_r = T_{mf}/T_c$	=	528/408	= 1.294
Bottom,	$T_r = T_{wf}/T_c$	=	540/408	= 1.324
Wellhead,	$p_r = p_{tf}/p_c$	=	1000/667	= 1.499

From equation (3.7)

$$F = \frac{(0.10797)(2.50)}{(2.441)^{2.612}} = 0.02624$$

$$F^2 = 0.000688$$

Left-hand side of equations (3.9) and (3.10)

$$37.5\ G\ \frac{d}{2}\ =\ (37.5)\ (0.74)\ (5000)/2\ =\ 69375.$$

Calculate l_{tf}:
The compressibility factor (see appendix D for calculations) for a reduced temperature and pressure of 1.262 and 1.499, respectively is

$$Z_{tf}\ =\ 0.690,\ p_{tf}/T_{tf}Z_{tf}\ =\ (1000)\ (515)\ (0.690)\ =\ 2.814$$

$$l_{tf}\ =\ \frac{(p_{tf}/T_{tf}\ Z_{tf})}{F^2\ +\ \dfrac{(p_{tf}/T_{tf}\ Z_{tf})^2}{1000}}\ =\ \frac{(2.814)}{(0.000688)\ +\ \dfrac{(2.814)^2}{1000}}\ =\ 326.96$$

Step 1 (the upper half of the flow string)
First trial

Assume

$$l_{mf}\ =\ l_{tf}\ =\ 326.96$$

Solving equation (3.9) for p_{mf}

$$69375.\ =\ (p_{mf}\ -\ 1000)\ (326.96\ +\ 326.96)$$

$$p_{mf}\ =\ 1106\ \text{psia}$$

Second trial

$$p_r\ =\ p_{mf}/p_c\ =\ 1106/667\ =\ 1.658$$

$$Z_{mf}\ =\ 0.680\ \text{at}\ T_r\ =\ 1.293\ (T_{mf}\ =\ 527.5°\text{R}),\ p_r\ =\ 1.658$$

$$(p_{mf}/T_{mf}\ Z_{mf})\ =\ (1106)/(527.5)\ (0.680)\ =\ 3.083$$

$$l_{mf}\ =\ 3.083/[.000688\ +\ (3.083)^2/1000]\ =\ 302.47$$

Solving equation (3.9) for p_{mf}

$$69375\ =\ (p_{mf}\ -\ 1000)\ (302.47\ +\ 326.96)$$

p_{mf} = 1110 psia

Third Trial

$p_r = p_{mf}/p_c = 1110/667 = 1.664$

$z_{mf} = 0.680$ at $T_r = 1.293$, $p_r = 1.664$

$(p_{mf}/T_{mf} Z_{mf}) = (1110)/(527.5) (0.680) = 3.095$

$I_{mf} = 3.095/(.000688 + (3.095)^2/1000) = 301.45$

Solving equation (3.9) for p_{mf}

$69375 = (p_{mf} - 1000) (301.45 + 326.96)$

p_{mf} = 1110 psia

Step 2 (the lower half of the flow string)
First trial

Assume

$I_{wf} = I_{mf} = 301.45$

Solving equation (3.10) for p_{wf}

$69375 = (p_{wf} - 1110) (301.45 + 301.45)$

P_{wf} = 1225 psia

Second trial

$P_r = p_{wf}/p_c = 1225/667 = 1.837$

$Z_{wf} = 0.693$ at $T_r = 1.324$ (at $T_{wf} = 540°R$), $p_r = 1.837$

$(p_{wf}/T_{wf} Z_{wf}) = (1225)/(540) (0.693) = 3.273$

$I_{wf} = 3.273/(0.000688 + (3.273)^2/1000) = 287.09$

Solving equation (3.10) for p_{wf}

$$69375 = (p_{wf} - 1110)(287.09 + 301.45)$$

$$p_{wf} = 1228 \text{ psia}$$

Third trial

$$p_r = p_{wf}/p_c = 1228/667 = 1.839$$

$$Z_{wf} = 0.693 \text{ at } T_r = 1.324, p_r = 1.839$$

$$(p_{wf}/T_{wf}Z_{wf}) = (1228)/(540)(0.693) = 3.281$$

$$I_{wf} = 3.281/[0.000688 + (3.281)^2/1000] = 286.48$$

Solving equation (3.10) for p_{wf}

$$69375 = (p_{wf} - 1110)(286.48 + 301.45)$$

$$P_{wf} = 1228 \text{ psia}$$

Parabolic Interpolation
From equation (3.11)

$$(69337. \times 2) = \frac{p_{wf} - p_{tf}}{3}[326.96 + 4(301.45) + 286.48]$$

$$p_{wf} - p_{tf} = 229$$

$$p_{wf} = 1000 + 299 = 1229 \text{ psia}$$

3.2 Stabilization Time and Radius of Investigation

In conducting the conventional test and the extended rates of flow for the isochronal tests, the duration of flow must be such as to reach stabilized conditions.

In practice, the time required to reach stabilization is usually considered to be the time to the onset of pseudo-steady-state flow. Earlier in this manual, an expression (see equation 2.20) was presented to estimate the onset of pseudosteady-state flow in a circular reservoir. It is important to recognize that stabilization time can be considerably longer than indicated by equation (2.20) when the geometrical configuration is not symmetrical. Also, the time to stabilization will be long for a large reservoir with low permeability and short for a limited reservoir of high permeability. It is always good practice to make an estimate of stabilization time before a test is begun. If a short stabilization time is expected, a conventional test may be run; but if stabilization time is expected to be long, isochronal or modified isochronal tests must be run. State regulatory bodies often define rule-of-thumb estimates of stabilization time of the form "less than 0.1% pressure drop in 15 minutes." It should be recognized that such rule-of-thumb estimates may be in error.

Another approach to estimating the time to stabilization may be made by defining a radius of investigation. The radius of investigation is defined as a circular region of the reservoir the pressure of which is affected by the flowing well located at its center. As time increases, more of the reservoir is influenced until the pressure transient reaches the outer boundary of the reservoir. At this point, the radius of investigation is equal to the external radius of the system and it no longer changes. Actually, this is the time when all the boundaries are felt, the pseudosteady-state flow period begins and stabilization is said to have been attained.

3.3
Choice of Flow Rates and Duration of Flow Rates

The pressure response during flow is dependent upon flow rates, therefore, maximum care must be given to the selection of the flow rates. In conducting any flow test, the minimum flow rate used should be at least equal to that required to lift the fluids, if any, from the well. The rate should also be sufficient to maintain a wellhead temperature above that which will allow hydrates to form. The suggested minimum and maximum test flow rates are those that result from pressure drawdowns of 5% and 25%, respectively, of the shut-in pressure. This is approximately equivalent to 10% and 75% of the AOF. The flow rates must not be so high as to cone water into the well from a gas-water contact nor to cause the production of sand. Also, the selected rates should not allow retrograde condensation to take place in the vicinity of the wellbore or in the wellbore itself.

A sequence of increasing flow rates is the most widely practiced procedure in deliverability tests. A decreasing sequence may be advisable if there is a possibility of hydrate formation because it will result in higher wellbore temperatures and this, in turn, will avoid hydrate formation. On the other hand, if liquid holdup in the wellbore is a problem, a decreasing sequence may be required.

Conventional and isochronal tests require the stabilization of pressure during shut-in periods. Therefore, as long as this requirement is fully satisfied, the rate sequence is immaterial. However, for the modified isochronal test, an increasing rate sequence should be used.

The duration of a flow period is determined by the radius of investigation (time to stabilization), as explained earlier and by the extent of wellbore storage effects—a topic to be discussed in the next section.

3.4 Wellbore Storage Effects

When a producing well is shut-in at the surface, flow from the formation into the wellbore continues for a period of time. This continuation of flow is called afterflow or wellbore storage and is caused by the fluid compressibility.

The wellbore storage time is given by the expression:

$$t_{ws} = \frac{36177 \, \mu \, V_{ws} \, C_{ws}}{kh} \qquad (3.11)$$

Where:

V_{ws} = volume of wellbore tubing (and annulus if there is no packer), or well depth times cross-sectional area of flow, cu ft

C_{ws} = compressibility of wellbore fluids, psi^{-1}

μ = viscosity of wellbore fluids, cp

k = permeability, md

h = formation thickness, ft

t_{ws} = wellbore storage time, hours

The wellbore storage time gives the approximate time in hours required for the wellbore storage effects to become negligible. Any pressure buildup data, then, must be recorded for a period well beyond the wellbore storage time if the data are to represent formation properties. The above equation indicates that wellbore storage effects are directly proportional to well depth and fluid viscosity and inversely proportional to the capacity of the formation (kh product). Also, wellbore storage effects decrease with increasing pressure because the compressibility of the fluid decreases as the pressure level is increased.

3.5 Production of Liquids

Many gas wells produce liquid (condensate and/or water), which cause fluctuations in rate and pressure measurements. In some cases, long flow times may be needed before the liquid-to-gas ratio stabilizes and test data become valid. Deliverability tests are valid even when this ratio becomes large. It is important though that a proper accounting of liquid production be made.

When liquid is produced, bottomhole pressure calculations using surface measurements are less reliable. It is suggested that, if significant amounts of liquids are being produced during testing, bottomhole pressures and temperatures should be measured directly.

Also separator systems should be employed to separate the liquid before metering the dry gas. The produced condensate, if any, should then be converted to its gas equivalent using the relationship:

$$G.E. = 133,000 \ \frac{\partial_{cond}}{M_{cond}} \qquad (3.12)$$

Where:

G. E. = gas equivalent, SCF/STB

∂_{cond} = specific gravity of condensate
            (water = 1.00)
M_{cond} = molecular weight of condensate

Where molecular weight of condensate can be estimated by:

$$M_{cond} \simeq \frac{44.29 - \partial_{cond}}{1.03 - \partial_{cond}} \qquad (3.13)$$

The total gas produced as gas equivalent is then added to the dry gas production to yield the total gas flowing from the reservoir during the test.

Since the reservoir may contain some liquid fraction, care must be taken to ensure that both the gas and produced fluids are properly separated and accurately measured.

The condensate and dry-gas volumes are recombined in correct proportions to estimate the specific gravity of the produced fluid. This specific gravity in turn can be used to estimate the critical temperature and pressure of the fluid which, in turn, is used in calculating z-factors. The specific gravity may be calculated as follows:

$$\partial_w = \frac{RG + 4584\,\partial_{cond}}{R + 132{,}800\,\dfrac{\partial_{cond}}{M_{cond}}} \qquad (3.14)$$

Where:

G = specific gravity of gas (air = 1.00)
∂_w = specific gravity of well fluid (air = 1.00)
R = ratio of dry gas to condensate, SCF/STB
∂_{cond} = specific gravity of condensate
            (water = 1.00)
M_{cond} = molecular weight of condensate

3.6 Single-Point Test

There are two situations where only a single-point test on a well is required: (1) where a regulatory agency fixes the value for n in the conventional test and a single data point is needed to obtain values for AOF and C (see, for example, the regulations of Oklahoma in chapter 4) and (2) where a single-point test is used in the updating of data obtained in a previous test. If earlier tests have provided values for n and the non-Darcy flow coefficient, D, then a single-point test is sufficient to update values of C and S. This test is often conducted annually as part of a pressure survey. In such cases, the gas is usually flowed directly into the pipeline and is not wasted. During the test, a single point on the deliverability curve is obtained.

3.7 The Orifice Meter

3.7.1 The Open System—Critical Flow Prover

During well testing, measurement of gas flow rates is usually determined by measuring the pressure drop across an orifice (restriction) placed in the pipeline. In the open-orifice metering system, referred to as the critical flow prover shown in figure 3.1, gas flows directly to the atmosphere. This type of gas metering is quick and easy for well testing; however, when the gas is vented, large pressure drops across the orifice may cause hydrates or ice to form.

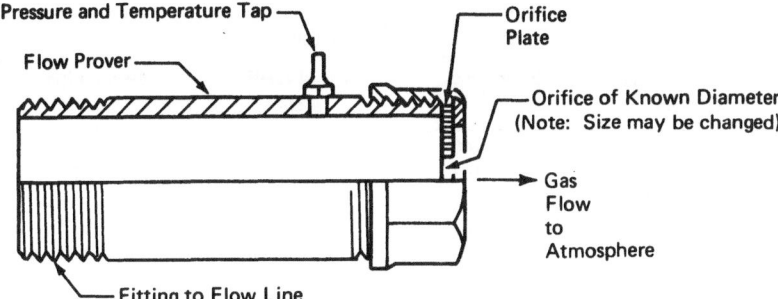

Pressure and Temperature Tap

Flow Prover

Orifice Plate

Orifice of Known Diameter
(Note: Size may be changed)

Gas
Flow
to
Atmosphere

Fitting to Flow Line

Figure 3.1
Schematic of Critical Flow Prover.

The flow rate through a criticial flow prover is given by the expression:

$$q_{sc} = 0.001 \, (C) \, (p) \, (F_{tf}) \, (F_g) \, (F_{pv}) \qquad (3.15)$$

Where:

q_{sc} = flow rate, MMSCF/D

p = flowing upstream pressure, psia

F_{tf} = flowing temperature factor $= \left(\dfrac{520}{T}\right)^{\frac{1}{2}}$

T = flowing temperature, °R

F_g = specific gravity factor $= \left(\dfrac{0.6}{G}\right)^{\frac{1}{2}}$

G = gas specific gravity, (air = 1.00)

F_{pv} = supercompressibility factor $= \left(\dfrac{1}{z}\right)^{\frac{1}{2}}$

z = supercompressibility at T and P.

C = orifice coefficient (see table 3.1)

In equation (3.15) the flowing temperature, flowing pressure and gas gravity are the quantities to be measured and values for C are to be found in table 3.1.

Table 3.1 Orifice Coefficients for 2-Inch and 4-Inch Critical Flow Provers. (From Railroad Commission of Texas 1950.)

Size of Orifice (inches)	2-Inch Pipe C	Size of Orifice (inches)	4-Inch Pipe C
1/16	0.0846	1/4	1.384
3/32	0.1863	3/8	3.110
1/8	0.3499	1/2	5.564
3/16	0.8035	5/8	8.668
7/32	1.1090	3/4	12.422
1/4	1.4360	7/8	16.893
5/16	2.2080	1	22.007
3/8	3.1420	1 1/8	27.721
7/16	4.5030	1 1/4	34.229
1/2	5.6530	1 3/8	41.210
5/8	8.5500	1 1/2	49.106
3/4	12.4900	1 3/4	67.082
7/8	17.1800	2	88.628
1	22.5800	2 1/4	113.617
1 1/8	28.9200	2 1/2	142.490
1 1/4	36.5100	2 3/4	176.420
1 3/8	44.8600	3	216.790
1 1/2	55.6400		

3.7.2 The Closed System—Orifice Meter

If environmental constraints prohibit gas venting to the atmosphere, gas should be metered through a closed-orifice system such as that shown in figure 3.2. Once again, the pressure drop is measured across an in-line orifice. The most common configuration, flange taps, is shown on the left in the figure.

The rate of flow of gas through a closed-orifice metering system is determined using the expression

$$q_{sc} = 24 \times 10^{-6} \, C \, (h_w p)^{1/2} \qquad (3.16)$$

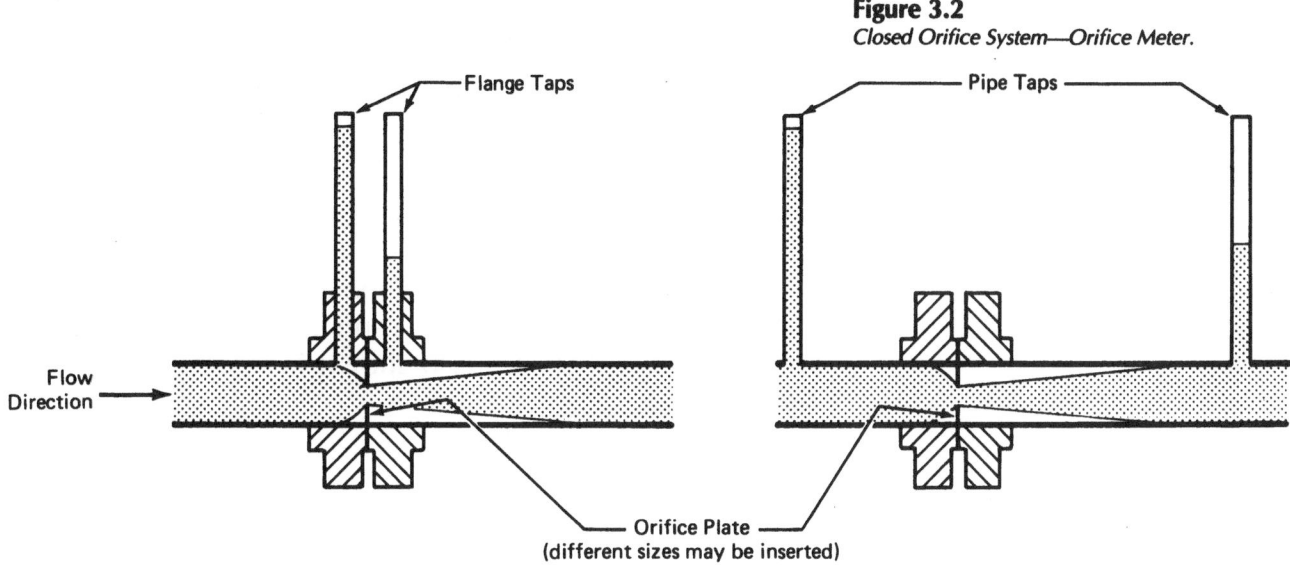

Figure 3.2
Closed Orifice System—Orifice Meter.

Flange Taps

Pipe Taps

Flow Direction →

Orifice Plate
(different sizes may be inserted)

Where:

q_{sc} = flow rate, MMSCF/D

p = static line pressure (usually the upstream pressure), psia

h_w = differential pressure across orifice, inches of H_2O at 60°F

C = orifice coefficient $(F_b)\ (F_r)\ (Y)\ (F_{pb})\ (F_{tb})\ (F_{tf})\ (F_g)$ (F_{pv})

F_b = basic orifice factor, (see Appendix E)

F_r = Reynold's number factor, (see Appendix E)

Y = expansion factor, (see Appendix E)

F_{pb} = pressure base factor = $14.73/p_b$

p_b = base pressure, psia

F_{tb} = temperature base factor = $(T_b + 460)/520$

T_b = base temperature, °F

F_{tf} = flowing temperature factor = $\left(\dfrac{520}{T}\right)^{1/2}$

T = flowing temperature, °R

$$F_g = \text{specific gravity factor} = \left(\frac{0.6}{G}\right)^{1/2}$$

$$G = \text{gas specific gravity (air} = 1.00)$$

$$F_{pv} = \text{supercompressibility factor} = \left(\frac{1}{z}\right)^{1/2}$$

$$z = \text{supercompressibility at T and p}$$

In designing a closed-orifice metering system, minimum lengths of straight pipe before and after the orifice must be established. These are also shown in figures included in Appendix E.

3.7.3
Turbine Meters

Turbine gas meters are now accepted in many jurisdictions for gaswell testing applications. The principle of operation of a turbine meter is based on the presence of a velocity sensing device within the meter. The velocity of flow through the meter is parallel to a rotor axis, and the speed of rotation is nominally proportional to the rate of flow. Flow rates, then, are converted to volumes and a readout device is mounted directly on the face plate of a gear housing. A photograph of a Rockwell TP Turbo-Meter is shown in figure 3.3. The volumetric flow rate measured by the meter depends on the gas gravity, temperature, pressure, and size of the turbine meter. The volumes recorded on the meter, then, must be corrected to standard temperature and pressure. Performance data for two sizes of Rockwell Turbo-Meters (2-inches and 3-inches) are shown in table 3.2.

Although turbine meters may have a high level of inaccuracy at low velocities, they otherwise have a number of advantages including the fact that they provide a direct readout of volume at line conditions.

Table 3.2 Performance Data—Rockwell Turbo-meters.

INLET PRESSURE PSIG	2″ — TP-4			3″ — TP-9		
	MAXIMUM MSCF FLOW RATE PER		RANGE	MAXIMUM MSCF FLOW RATE PER		RANGE
	HOUR	DAY	±1%	HOUR	DAY	±1%
0.25	4	96	5:1	9	216	10:1
5	5.3	120	6:1	12	286	11:1
10	6.7	161	7:1	15	360	13:1
15	8.0	192	7:1	18	433	14:1
20	9.4	225	7.5:1	21	507	15:1
25	10.7	257	8:1	24	581	16:1
50	17.6	422	10:1	39	950	21:1
75	24.4	585	12:1	55	1318	25:1
100	31.2	749	14:1	70	1687	28:1
200	58.5	1404	19:1	132	3161	38:1
300	85.8	2059	23:1	193	4636	46:1
400	113.1	2714	26:1	255	6110	53:1
500	140.4	3370	30:1	316	7584	59:1
600	167.7	4025	33:1	377	9059	65:1
700	195.0	4680	35:1	439	10533	70:1
800	222.4	5337	37:1	500	12007	74:1
900	249.7	5993	40:1	562	13482	79:1
1000	277.0	6648	42:1	623	14956	83:1
1100	304.3	7303	45:1	685	16431	87:1
1200	331.6	7958	47:1	746	17905	91:1
1300	358.9	8613	48:1	807	19379	95:1
1400	386.2	9268	50:1	869	20854	98:1

NOTE: For 0.6 sp. gr. gas at 60°F and 14.73 psia.

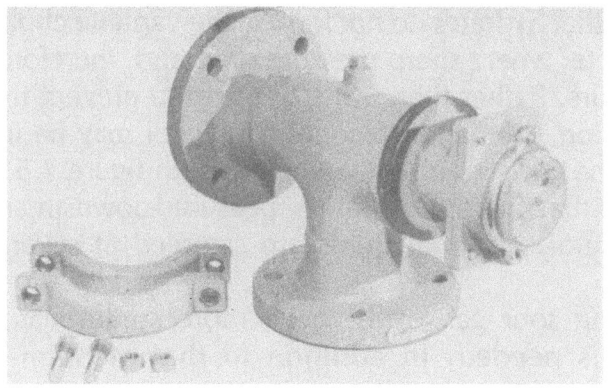

Figure 3.3
Photograph of a Rockwell TP Turbo-Meter.

3.8
Testing
Equipment

The selection of testing equipment is based on the nature of the produced fluids and the type of the test to be run.

The simplest configuration of wellhead testing facilities is that required for a well producing a sweet, dry gas. In this case, the essential equipment is a flow-rate measurement device, a shut-in and flowing pressure measurement device, a thermometer, gas sampling equipment, and necessary fittings. If a critical flow prover is used, the gas is vented to the atmosphere (figure 3.4) unless, of course, regulations specify otherwise. If the gas is to be produced into the gathering system, an orifice meter using a permanent or removable meter run is used.

The presence of condensate and/or water in the produced fluid requires somewhat more complex testing equipment. Under these conditions, in addition to the equipment needed for a sweet, dry gas, it is necessary to install gas and liquid sampling equipment, line heaters, separation facilities, and liquid measuring devices. If three phases are present, then two stages of separation are required. For high-pressure wells, more than one stage of separation may be necessary. The line heater is used to heat the fluids so that hydrates do not form at the variable choke or orifice meter where sharp pressure drop and, therefore, cooling occurs. Rather than using a heater to prevent hydrate formation, methanol, alcohol, or glycol may be injected into the gas stream at the wellhead. In figure 3.5 it will be noted that the gas flow rate is measured downstream of the separator, and that fluids are sampled at several points.

In testing sour gas wells, even more sophisticated equipment is needed. In addition to that mentioned

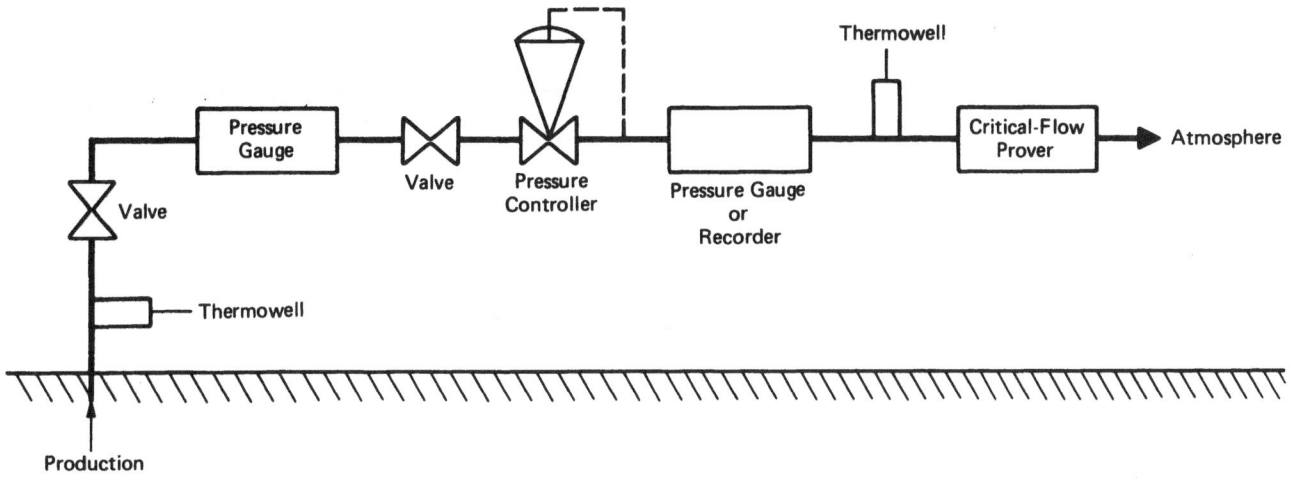

Figure 3.4
Wellhead Facilities for Testing Dry Gas Wells (with Critical Flow Prover).

previously, a flow line to an appropriate flare stack is required. In Alberta, Canada, for example, if the gas contains over 1% H_2S, it must be flared from a stack at least 12 meters high and flaring must take place at least 25 meters from the wellhead.* Liquid seals may also be needed to protect the gas meter and dead-weight tester from the corrosive nature of the H_2S gas. Other measures to protect against the corrosive nature of the gas might need to be taken.

---

*Remember that prolonged breathing of H_2S gas (which is extremely poisonous) can be fatal.

Figure 3.5
*Wellhead Testing Facilities for Testing Gas
Wells Producing Condensate and/or Water.*

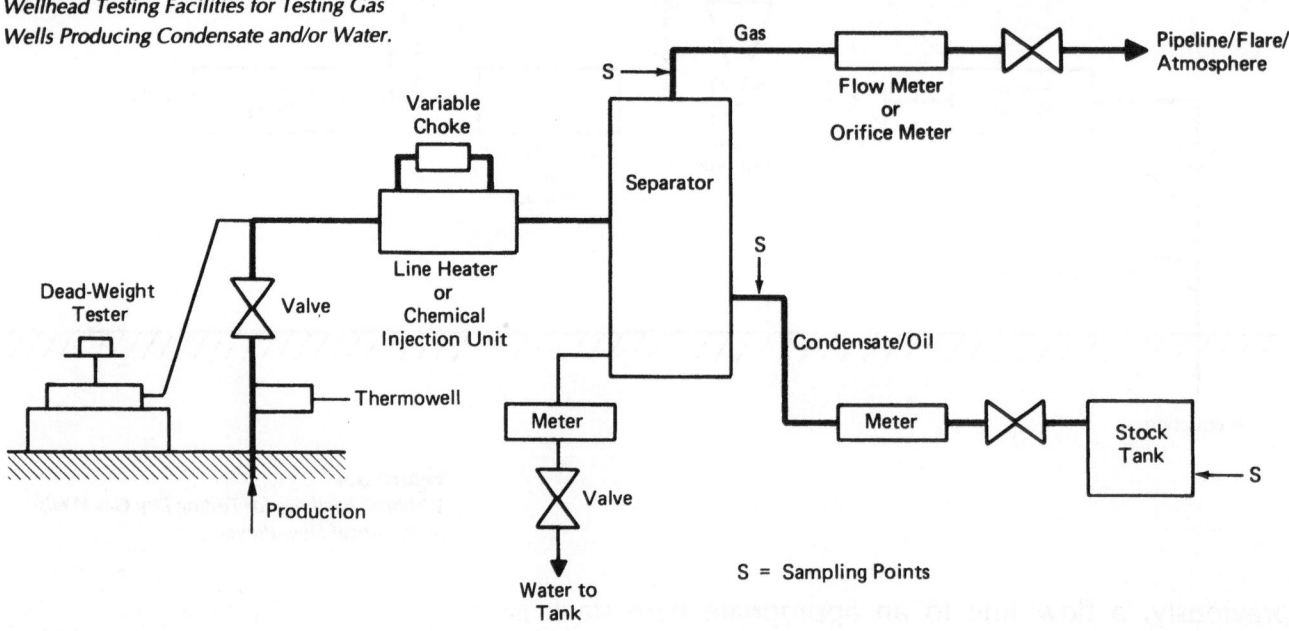

3.9
Measurement,
Sampling, and
Safety in
Testing

3.9.1
Volume Measurement

We mentioned earlier that gas flow rate measurement may be accomplished through the use of an orifice meter, a critical flow prover, or a turbine meter.

Turbine or displacement meters are generally used to

measure condensate flow rates. The meter should be installed upstream of a snap-acting valve and with sufficient length of straight pipe for an accurate measurement. The size of the separator and valve setting should be regulated so as to ensure sufficient retention time in the separator for the fluids to reach equilibrium. Gas breakout in the meter will cause errors in measurement. If condensate volumes are measured in the stock tank rather than by using a meter, it is important for accuracy and safety that the fluid entering the stock tank come from a low-pressure separator. For increased accuracy, efforts to recover and measure tank vapors should be made.

Water flow rates may be measured with a turbine or displacement meter or, alternatively, gauged in the tank. If both condensate and water are gauged in the same tank, then pink tape is used to show water level. If meters are used, a snap-acting valve should be used to ensure sufficient flow to activate the meter.

3.9.2
Pressure Measurement

Accurate measurement of pressures, both static and flowing, is of singular importance in gaswell testing. Our primary objective is the measurement of sandface pressures. Ideally, this pressure is measured downhole at the sandface or opposite the perforations using an accurate, carefully calibrated bottomhole pressure gauge. The three basic types of bottomhole gauges are the *self-contained wireline gauge*, the *permanently installed surface-recording gauge*, and the *retrievable surface-recording gauge*. The self-contained wireline gauge is the one most widely used in the industry. It is lowered into the hole on a solid wire or slick line. The three essential components of the gauge are: the pressure-sensing device, a pressure-time recording chart, and a clock. The clock is designed to run and record data on a chart for a specific period of time. Different clocks are available for different recording times. If data are needed beyond that time, the gauge must be retrieved, reset, and rerun into the well. A number of companies manufacture and sell these gauges. An Amerada gauge is shown in figure 3.6.

Permanently installed surface-recording gauges are generally attached to the tubing string. The instrument includes a means for measuring bottomhole pressure and transmitting that value to the surface. Normally, a cable, strapped to the outside of the production tubing, is used to transmit information. As data are needed, a recorder at the surface is connected and turned on.

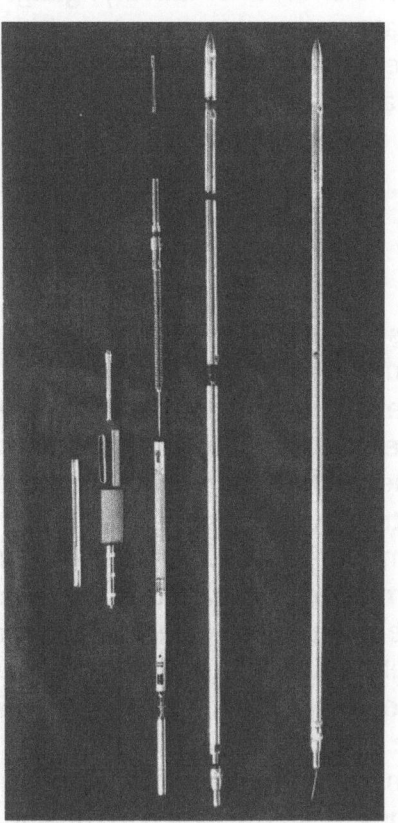

Figure 3.6
An Amerada RP—3 Gauge. (Courtesy Geophysical Research Corporation.)

Retrievable surface-recording gauges are similar in most respects to the permanently installed type except that they are run on a simple conductor wireline. Most use Bourden tubes for pressure measurement, although Hewlett-Packard uses a quartz crystal, which has a better resolution.

Even though bottomhole pressures are being recorded, it is recommended that wellhead pressure measurements be recorded at all times. Such data can be used as a check on bottomhole data and wellhead deliverability calculations. Surface pressures are measured continuously. In some instances, for example, when there are mechanical difficulties, highly deviated holes, or sour gas production, it may not be practical to run bottomhole gauges. In such situations, wellhead pressures are measured, and correlations mentioned at the beginning of this chapter are used to convert the wellhead pressures to the sandface pressures.

3.9.3 Sampling

All fluids produced at the wellhead should be accurately measured, sampled, and analyzed to assure that an accurate producing rate is calculated. Samples are taken following industry practice, and all sampling points and conditions of sampling should be noted on a line diagram of the wellhead facilities. Liquid samples should be taken at separator conditions so as to avoid the calculations required to convert stock tank volumes to equivalent separator conditions. Of course, reference should be made to more accurate sample data that might have been collected soon after discovery.

3.9.4
Safety

Safety enters into all aspects of oil and gas field operations. Gaswell testing is no exception. Some elements of safety, like the precautions to be taken during the flaring of gas, have already been discussed. All lines should be pressure tested before the test and tied down so they will not move during the test. It is often recommended that a test not be started after dark. Alberta recommends that a minimum five-person crew, (two two-person shifts and a supervisor), be assigned to any 24-hour testing period. It also requires that two sets of breathing apparatus be available on the lease at all times; and, of course, emergency procedures and fire fighting capabilities should be fully understood by all crew members before the test begins. Supervising engineers should take time to inform themselves of company practices and local regulatory standards before a test and be sure that they are fully observed. Careful observance of safety procedures by the engineer will prevent unnecessary injuries to workers, damage to the producing well, surface equipment, or landowner's property.

References

[1] Energy Resources Conservation Board: *Gas Well Testing: Theory and Practice,* fourth edition (metric), Alberta, Canada (1979) 4.

[2] Cullender, M.H. and Smith R.V.: "Practical Solution of Gas-Flow Equations for Well and Pipelines With Large Termperature Gradients," *Trans.,* AIME (1956) 207, 281–287.

[3] Nikuradse, J.: "Laws of Fluid Flow in Rough Pipes (Part One)," *Pet. Eng.* (March 1940) 164.

[4] Smith, R.V., Miller, J.S., and Ferguson, J.W.: *Flow of Natural Gas Through Experimental Pipe Lines and Transmission Lines,* Monograph 9, U.S. Bureau of Mines, 1956.

[5] Alberta Energy Resources Conservation Board, "The Oil and Gas Conservation Act and Oil and Gas Conservation Regulations," Alberta (July 1979) 85–86.

Chapter 4

Procedures & Regulations

The field procedures followed during a gaswell test are usually those specified by the regulatory agencies in the state or province in which the well is located. A company will often do more testing then that required by local regulations, but those regulations usually form the basis of the test procedures.

Many state regulatory agencies and operating companies refer to the *Manual of Back-Pressure Testing of Gas Wells* (Interstate Oil Compact Commission (IOCC), 1962) as their primary guide to gaswell testing. Individual states have published their own respective backpressure test procedures. The procedures recommended by the IOCC and the key elements of the regulations of Texas, Oklahoma, Louisiana, New Mexico, and Alberta have been reproduced below.

Test data and calculations should be submitted to the appropriate regulatory agency on standard reporting forms. Each agency has its own set of reporting forms. A model set is included in Appendix F. This set is substantially similar to those issued by the Province of Alberta. An example of the reporting of a conventional backpressure test is also shown in Appendix F.

4.1
Interstate Oil Compact Commission Procedures*

RULES OF PROCEDURE

I.
General Instructions

All backpressure tests required by the (*State Regulatory Body*) shall be conducted in accordance with the procedures set out below except for those wells in pools where special testing procedures are applicable.

Calculations shall be made in the manner prescribed in the appropriate test examples. The observed data and calculations shall be reported on the prescribed forms.

Gas produced from wells connected to a gas transportation facility should not be vented to the atmosphere during testing. When an accurate test can be obtained only under conditions requiring venting, the volume shall be the minimum required to obtain an accurate test.

All surface pressure readings shall be taken with a dead weight gauge. Under special conditions where the use of a dead weight gauge is not practical, a properly calibrated spring gauge may be used when authorized by the (*State Regulatory Body*). Subsurface pressures determined by the use of a properly calibrated pressure bomb are acceptable.

The temperature of the gas column must be accurately known to obtain correct test results, therefore a thermometer well should be installed in the wellhead. Under shut-in or low flow rate conditions, the observed wellhead temperatures may be distorted by the external temperature. Whenever this situation exists the mean annual temperature should be used.

---

*IOCC Procedures, Box 53127, Oklahoma City, OK 73105

II.
Multipoint Back-Pressure Test Procedures

A multipoint back-pressure test shall be taken for the purpose of determining the absolute open flow and exponent n from the plot of the equation

$$Q = C(P_f^2 - P_s^2)^n \qquad \text{(III–1)}$$

or, under certain conditions as prescribed herein,

$$Q = C(P_c^2 - P_w^2)^n \qquad \text{(III–2)}$$

A. Stabilized Multipoint Test

1. Shut-in Pressure

a. Wells with a pipeline connection shall be produced for a sufficient length of time at a flow rate large enough to clear the well bore of accumulated liquids prior to the shut-in period. If the well bore cannot be cleared of accumulated liquids while producing into a pipeline, the well shall be blown to the atmosphere to remove these liquids.

b. Wells without pipeline connections shall be blown to the atmosphere to remove accumulated liquids.

c. The well shall be shut in until the rate of pressure buildup is less than 1/10 of 1 per cent of the previously recorded pressure, psig, in 30 minutes. This pressure shall be recorded.

2. Flow Tests

a. After recording the shut-in pressure, a series of at least four stabilized flow rates and the pressures corresponding to each flow rate shall be taken. Any shut-in time between flow rates shall be held to a minimum. These rates shall be run in the increasing flow-rate sequence. In the case of high liquid ratio wells or unusual temperature conditions, a decreasing flow-rate sequence may be used if the increasing sequence method did not result in point alignment. If the decreasing sequence method is used, a statement giving the reasons why the use of such method was necessary, together with a copy of the data taken by the increasing sequence method, shall be furnished to the (*State Regulatory Body*). If experience has shown that the use of the decreasing sequence method is necessary for an accurate test, a test by the increasing sequence method will not be required.

b. The lowest flow rate shall be a rate sufficient to keep the well clear of all liquids.

c. One criterion as to the acceptability of the test is a good spread of data points. In order to assure a good spread of points, the wellhead flowing pressure, psig, at the lowest flow rate should not be more than 95 per cent of the well's shut-in pressure, psig, and at the highest flow rate not more than 75 per cent of the well's shut-in pressure, psig. If data cannot be obtained in accordance with the foregoing provisions, an explanation shall be furnished to the (*State Regulatory Body*).

d. All flow-rate measurements shall be obtained by the use of an orifice meter, critical flow prover, positive choke, or other authorized metering device in good operating condition. When an orifice meter is used as the metering device, the meter shall be calibrated and the diameters of the orifice plate and meter run verified as to size, condition, and compliance with acceptable standards. The differential pen shall be zeroed under operating pressure before beginning the test.

e. The field barometric pressure shall be determined.

f. The specific gravity of the separator gas and of the produced liquid shall be determined.

g. Periodically, during each flow rate, wellhead flowing pressures and flow-rate data shall be recorded and liquid production rate shall be observed to aid in determining when stabilization has been attained. A constant flowing wellhead pressure or static column wellhead pressure and rate of flow for a period of at least 15 minutes shall constitute stabilization for this test.

h. At the end of each flow rate the following information shall be recorded: (1) Flowing wellhead pressure. (2) Static column wellhead pressure if it can be obtained. (3) Rate of liquid production. (4) Flowing wellhead temperature. (5) All data pertinent to the gas metering device.

3. Calculations

a. General

A wellhead absolute open flow as determined from the wellhead equation, (III–2), $Q = C(P_c^2 - P_w^2)^n$, is normally found to be equivalent to the bottom-hole absolute open flow as determined from the bottom-hole equation, (III-1), $Q = C(P_f^2 - P_s^2)^n$, where the wellhead shut-in pressure of all wells in a given reservoir is below 2000 psig. Under this condition the wellhead absolute open flow is acceptable instead of the bottom-hole absolute open flow.

b. Bottom-hole Calculations

(1) Bottom-hole pressures shall be calculated to a datum at the mid-point of the production section open to flow. The point of entry into the tubing may be used as the datum if it is not more than 100 feet above or below the mid-point of the producing section open to flow.

(2) Under all shut-in conditions and under flowing conditions, when the static column wellhead pressures can be obtained, the

bottom-hole pressure shall be calculated as shown in test examples 1 and 2.

    (3) When only the flowing wellhead pressures can be obtained, the bottom-hole pressures shall be calculated as shown in test example 3.

    (4) When the bottom-hole pressures are recorded by use of a properly calibrated bottom-hole pressure bomb and corrected to the proper datum, these pressures may be used in the bottom-hole formula.

    (5) When liquid accumulation in the well bore during the shut-in period appreciably affects the wellhead shut-in pressure, the calculation of the bottom-hole pressure shall be made as shown in test example 8.

    c. Wellhead Calculations

    (1) The static column wellhead pressure must be obtained if possible.

    (2) When only the flowing wellhead pressures can be obtained, the static column wellhead pressures shall be calculated as shown in test example 3.

    (3) When liquid accumulation in the well bore during the shut-in period appreciably affects the wellhead shut-in pressure, appropriate correction of the surface pressure shall be made. This correction shall be made in the manner shown in test example 8 or, at the option of the operator, by using a bottom-hole pressure bomb and correcting to wellhead conditions as shown in test example 9 or 10.

4. Reports

Upon completion of the test, all calculations shall be shown on Form BPT 5 and, if applicable, Form BPT 2, BPT 3, or BPT 4. ( ) copies of these forms and the back pressure curve described below shall be submitted to the (*State Regulatory Body*).

5. Plotting

    a. The points for the back-pressure curve shall be accurately and neatly plotted on equal scale log-log paper (3-inch cycles are recommended) and a straight line drawn through the best average of three or more points. When no reasonable relationship can be established between three or more points, the well shall be retested.

    b. The cotangent of the angle this line makes with the volume coordinate is the exponent n which is used in the back-pressure equation (III–1 or III–2). The exponent n shall always be calculated as shown in basic calculation 5.

    c. If the exponent n is greater than 1.000 or less than 0.500, the well shall be retested.

    d. If, after retesting the well, no reasonable alignment is established between three or more points, then a straight line shall be drawn through the best average of at least three points *of the retest.*

(1) If the exponent *n* is greater than 1.000, a straight line with an exponent *n* of 1.000 shall be drawn through the point corresponding to the highest flow rate utilized in establishing the line.

(2) If the exponent *n* is less than 0.500, a straight line with an exponent *n* of 0.500 shall be drawn through the point corresponding to the lowest flow rate utilized in establishing the line.

B. Non-Stabilized Multipoint Tests

When well stabilization is impractical to obtain for a series of points, or when gas must be flared during the test, the exponent *n* of the back-pressure curve shall be established by either the Constant Time Multipoint Test or the Isochronal Multipoint Test. The exponent *n* so determined shall then be applied to a stabilized one-point test to determine the absolute open flow. (See *Stabilized One-Point Back-Pressure Test Procedure.*) The flow during this one-point test shall be for a period adequate to reach stabilized conditions unless determination is made in conjunction with the (*State Regulatory Body*) that it would be impractical to continue flow until complete stabilization is reached.

1. Constant Time Multipoint Test

a. Shut-in Pressure

(1) Wells with a pipeline connection shall be produced for a sufficient length of time at a flow rate large enough to clear the well bore of accumulated liquids prior to the shut-in period. If the well bore cannot be cleared of accumulated liquids while producing into a pipeline, the well shall be blown to the atmosphere to remove these liquids.

(2) Wells without pipeline connections shall be blown to the atmosphere to remove accumulated liquids.

(3) The well shall be shut in until the rate of pressure buildup is less than 1/10 of 1 per cent of the previously recorded pressure, psig, in 30 minutes. This pressure shall be recorded.

b. Flow Tests

(1) After recording the shut-in pressure, a series of at least four flow rates of the same duration and the pressures corresponding to each flow rate shall be taken. Any shut-in time between flow rates shall be held to a minimum. These rates shall be run in the increasing flow-rate sequence. In the case of high liquid ratio wells or unusual temperature conditions, a decreasing flow-rate sequence may be used if the increasing sequence method did not result in the alignment of points. If the decreasing sequence method is used, a statement giving the reasons why the use of such method was necessary, together with a copy of the

data taken by the increasing sequence method, shall be furnished the (*State Regulatory Body*). If experience has shown that the use of the decreasing sequence method is necessary for an accurate test, a test by the increasing sequence method will not be required.

(2) The lowest flow rate shall be a rate sufficient to keep the well clear of all liquids.

(3) One criterion as to the acceptability of the test is a good spread of data points. In order to assure a good spread of points, the wellhead flowing pressure, psig, at the lowest flow rate should not be more than 95 per cent of the well's shut-in pressure, psig, and at the highest flow rate not more than 75 per cent of the well's shut-in pressure, psig. If data cannot be obtained in accordance with the foregoing provisions, an explanation shall be furnished the (*State Regulatory Body*).

(4) All flow rate measurements shall be obtained by the use of an orifice meter, critical flow prover, positive choke, or other authorized metering device in good operating condition. When an orifice meter is used as the metering device, the meter shall be calibrated and the diameters of the orifice plate and meter run verified as to size, condition, and compliance with acceptable standards. The differential pen shall be zeroed under operating pressure before beginning the test.

(5) The field barometric pressure shall be determined.

(6) The specific gravity of the separator gas and of the produced liquid shall be determined.

(7) At the end of each flow rate the following information shall be recorded: (a) Flowing wellhead pressure. (b) Static column wellhead pressure if it can be obtained. (c) Rate of liquid production. (d) Flowing wellhead temperature. (e) All data pertinent to the gas metering device.

(8) The stabilized one-point test data may be obtained by continuation of the last flow rate in the manner prescribed for Flow Test in the *Stabilized One-Point Back-Pressure Test Procedure.*

c. Calculations

(1) General—A wellhead absolute open flow as determined from the wellhead equation, (III–2), $Q = C (P_c^2 - P_w^2)^n$, is normally found to be equivalent to the bottom-hole absolute open flow as determined from the bottom-hole equation, (III–1), $Q = C(P_f^2 - p_s^2)^n$, when the wellhead shut-in pressure of all wells in a given reservoir is below 2000 psig. Under this condition the wellhead absolute open flow is acceptable instead of the bottom-hole absolute open flow.

(2) Bottom-Hole Calculations

(a) Bottom-hole pressures shall be calculated to a datum at the mid-point of the producing section open to flow. The point of entry into the tubing may be used as the datum if it is not more than 100 feet above or below the mid-point of the producing section open to flow.

(b) Under all shut-in conditions and under flowing conditions, when the static column wellhead pressures can be obtained, the bottom-hole pressures shall be calculated as shown in test example 1 and 2.

(c) When only the flowing wellhead pressures can be obtained, the bottom-hole pressures shall be calculated as shown in test example 3.

(d) When the bottom-hole pressures are recorded by use of a properly calibrated bottom-hole pressure bomb and corrected to the proper datum, these pressures may be used in the bottom-hole formula.

(e) When liquid accumulation in the well bore during the shut-in period appreciably affects the wellhead shut-in pressure, the calculation of the bottom-hole pressure shall be made as shown in test example 8.

(3) Wellhead Calculations

(a) The static column wellhead pressure must be obtained if possible.

(b) When only the flowing wellhead pressures can be obtained, the static column wellhead pressures shall be calculated as shown in test example 3.

(c) When liquid accumulation in the well bore during the shut-in period appreciably affects the wellhead shut-in pressure, appropriate correction of the surface pressure shall be made. This correction shall be made in the manner shown in test example 8 or, at the option of the operator, by using a bottom-hole pressure bomb and correcting to wellhead conditions as shown in test example 9 or 10.

d. Reports—Upon completion of the test, all calculations shall be shown on Form BPT 5 and, if applicable, Form BPT 2, BPT 3, or BPT 4. ( ) copies of these forms and the back-pressure curve described below shall be submitted to the (*State Regulatory Body*).

e. Plotting

(1) The points for the back-pressure curve shall be accurately and neatly plotted on equal scale log-log paper (3-inch cycles are recommended) and a straight line drawn through the best average of three or more points. When no reasonable relationship can be established between three or more points, the well shall be retested.

(2) The cotangent of the angle this line makes with the volume

coordinate is the exponent n which is used in the back-pressure equation (III–1 or III–2). The exponent n shall always be calculated as shown in basic calculation 5.

(3) If the exponent n is greater than 1.000 or less than 0.500, the well shall be retested.

(4) If, after retesting the well, no reasonable alignment is established between three or more points, then a straight line shall be drawn through the best average of at least three points *of the retest.*

(a) If the exponent n is greater than 1.000, a straight line with an exponent n of 1.000 shall be drawn through the point corresponding to the highest rate of flow utilized in establishing the line.

(b) If the exponent n is less than 0.500, a straight line with an exponent n of 0.500 shall be drawn through the point corresponding to the lowest rate of flow utilized in establishing the line.

(5) The constant time data points are used only to determine the value of the exponent n. The back-pressure curve shall be drawn through the stabilized data point and parallel to the line established by the constant time data points. The absolute open flow may be determined from this back-pressure curve or calculated as shown in test example 4.

2. Isochronal Multipoint Test

   a. Shut-in Pressures

(1) Wells with a pipeline connection shall be produced for a sufficient length of time at a flow rate large enough to clear the well bore of accumulated liquids prior to the shut-in period. If the well bore cannot be cleared of accumulated liquids while producing into a pipeline, the well shall be blown to the atmosphere to remove these liquids.

(2) Wells without pipeline connections shall be blown to the atmosphere to remove accumulated liquids.

(3) Prior to each flow test as described below, the well shall be shut in until the rate of pressure buildup is less than 1/10 of 1 per cent of the previously recorded pressure, psig, in 30 minutes. This pressure shall be recorded and used with the data from the subsequent flow test.

   b. Flow Tests

(1) After recording the initial shut-in pressure, a series of at least four flow rates of the same duration and the pressures corresponding to each flow rate shall be taken. Each flow rate shall be preceded by a shut-in pressure as prescribed above in 2.a.(3).

(2) The lowest flow rate shall be a rate sufficient to keep the well clear of all liquids.

(3) One criterion as to the acceptability of the test is a good spread of data points. In order to assure a good spread of points, the wellhead flowing pressure, psig, at the lowest flow rate should not be more than 95 per cent of the well's shut-in pressure, psig, and at the highest flow rate not more than 75 per cent of the well's shut-in pressure, psig. If data cannot be obtained in accordance with the foregoing provisions, an explanation shall be furnished to the (*State Regulatory Body*).

(4) All flow rate measurements shall be obtained by the use of an orifice meter, critical flow prover, positive choke, or other authorized metering device in good operating condition. When an orifice meter is used as the metering device, the meter shall be calibrated and the diameters of the orifice plate and meter run verified as to size, condition, and compliance with acceptable standards. The differential pen shall be zeroed under operating pressure before beginning the test.

(5) The field barometric pressure shall be determined.

(6) The specific gravity of the separator gas and of the produced liquid shall be determined.

(7) At the end of each flow rate the following information shall be recorded: (a) Flowing wellhead pressure. (b) Static column wellhead pressure if it can be obtained. (c) Rate of liquid production. (d) Flowing wellhead temperature. (e) All data pertinent to the gas metering device.

(8) The stabilized one-point test data may be obtained by continuation of the last flow rate in the manner prescribed for Flow Test in the *Stabilized One-Point Back-Pressure Test Procedure*.

c. Calculations

(1) General

A wellhead absolute open flow as determined from the wellhead equation, (III–2), $Q = C(P_c^2 - P_w^2)^n$, is normally found to be equivalent to the bottom-hole absolute open flow as determined from the bottom-hole equation, (III–1), $Q = C(P_f^2 - P_s^2)^n$, where the wellhead shut-in pressure of all wells in a given reservoir is below 2000, psig. Under this condition the wellhead absolute open flow is acceptable instead of the bottomhole absolute open flow.

(2) Shut-in Pressure

The shut-in pressure preceding each flow rate shall be used in conjunction with the static column wellhead pressure corresponding to that flow rate.

(3) Bottom-hole Calculations

(a) Bottom-hole pressures shall be calculated to a datum at

the mid-point of the producing section open to flow. The point of entry into the tubing may be used as the datum if it is not more than 100 feet above or below the mid-point of the producing section open to flow.

(b) Under all shut-in conditions and under flowing conditions, when the static column wellhead pressures can be obtained, the bottom-hole pressures shall be calculated as shown in test examples 1 and 2.

(c) When only the flowing wellhead pressures can be obtained, the bottom-hole pressures shall be calculated as shown in test example 3.

(d) When the bottom-hole pressures are recorded by use of a properly calibrated bottom-hole pressure bomb and corrected to the proper datum, these pressures may be used in the bottom-hole formula.

(e) When liquid accumulation in the well bore during the shut-in period appreciably affects the wellhead shut-in pressure, the calculation of the bottom-hole pressure shall be made as shown in test example 8.

(4) Wellhead Calculations

(a) The static column wellhead pressure must be obtained if possible.

(b) When only the flowing wellhead pressures can be obtained, the static column wellhead pressures shall be calculated as shown in test example 3.

(c) When liquid accumulation in the well bore during the shut-in period appreciably affects the wellhead shut-in pressure, appropriate correction of the surface pressure shall be made. This correction shall be made in the manner shown in test example 8 or, at the option of the operator, by using a bottom-hole pressure bomb and correcting to wellhead conditions as shown in test example 9 or 10.

d. Reports

Upon completion of the test, all calculations shall be shown on Form BPT 5 and, if applicable, Form BPT 2, BPT 3, or BPT 4. ( ) copies of these forms and the back-pressure curve described below shall be submitted to the (*State Regulatory Body*).

e. Plotting

(1) The points for the back-pressure curve shall be accurately and neatly plotted on equal scale log-log paper (3-inch cycles are recommended) and a straight line drawn through the best average of three or more points. When no reasonable relationship can be established between three or more points, the well shall be retested.

(2) The cotangent of the angle this line makes with the volume

co-ordinate is the exponent n which is used in the back-pressure equation (III–1 or III–2). The exponent n shall always be calculated as shown in basic calculation 5.

(3) If the exponent n is greater than 1.000 or less than 0.500, the well shall be retested.

(4) If, after retesting the well, no reasonable alignment is established between three or more points, then a straight line shall be drawn through the best average of at least three points *of the retest.*

(a) If the exponent n is greater than 1.000, a straight line with an exponent n of 1.000 shall be drawn through the point corresponding to the highest rate of flow utilized in establishing the line.

(b) If the exponent n is less than 0.500, a straight line with an exponent n of 0.500 shall be drawn through the point corresponding to the lowest rate of flow utilized in establishing the line.

(5) The Isochronal data points are used only to determine the value of the exponent n. The back-pressure curve shall be drawn through the stabilized data point and parallel to the line established by the Isochronal data points. The absolute open flow may be determined from this back-pressure curve or calculated as shown in test example 4.

III.
Stabilized One-Point
Back-Pressure Test Procedure

The most recently determined exponent n established by a *Stabilized Multipoint Back-Pressure Test,* a *Constant Time Multipoint Test,* or an *Isochronal Multipoint Test* shall be used with this stabilized one-point test to determine the absolute open flow.

The Flow Test portion of this test may be used in conjunction with the *Constant Time Multipoint Test* or the *Isochronal Multipoint Test* to determine the absolute open flow.

A. DETERMINATION OF ABSOLUTE OPEN FLOW

1. Shut-in Pressure

a. Wells with a pipeline connection shall be produced for a sufficient length of time at a flow rate large enough to clear the well bore of accumulated liquids prior to the shut-in period. If the well bore cannot be cleared of accumulated liquids while producing into a pipeline, the well shall be blown to the atmosphere to remove these liquids.

b. Wells without pipeline connections shall be blown to the atmosphere to remove accumulated liquids.

c. The well shall be shut in until the rate of pressure buildup is less than 1/10 of 1 per cent of the previously recorded pressure, psig, in 30 minutes. This pressure shall be recorded.

2. Flow Test

a. After recording the shut-in pressure, a stabilized flow rate and its corresponding pressure shall be taken.

b. This flow rate shall be a rate sufficient to keep the well clear of all liquids.

c. The wellhead flowing pressure, psig, shall not be more than 95 per cent of the well's shut-in pressure, psig. If the data cannot be obtained in accordance with the foregoing provision, an explanation shall be furnished to the (State Regulatory Body).

d. The flow rate measurement shall be obtained by the use of an orifice meter, critical flow prover, positive choke, or other authorized metering device in good operating condition. When an orifice meter is used as the metering device, the meter shall be calibrated and the diameters of the orifice plate and meter run verified as to size, condition, and compliance with acceptable standards. The differential pen shall be zeroed under operating pressure before beginning the test.

e. The field barometric pressure shall be determined.

f. The specific gravity of the separator gas and of the produced liquid shall be determined.

g. Periodically during the flow rate, wellhead flowing pressures and flow-rate data shall be recorded and liquid production rate shall be observed to aid in determining when stabilization has been attained. A constant flowing wellhead pressure or static column wellhead pressure and rate of flow for a period of at least 15 minutes shall constitute stabilization for this test.

h. At the end of the flow rate the following information shall be recorded: (1) Flowing wellhead pressure. (2) Static column wellhead pressure if it can be obtained. (3) Rate of liquid production. (4) Flowing wellhead temperature. (5) All data pertinent to the gas metering device.

3. Calculations

a. General

A wellhead absolute open flow is determined from the wellhead equation, (III–2), $Q = C (P_c^2 - P_w^2)^n$, is normally found to be equivalent to the bottom-hole absolute open flow as determined from the bottom-hole equation, (III–1), $Q = C (P_f^2 - P_s^2)^n$, where the wellhead shut-in pressure of all wells in a given reservoir is below

2000 psig. Under this condition the wellhead absolute open flow is acceptable instead of the bottom-hole absolute open flow.

 b. Bottom-hole Calculations

(1) Bottom-hole pressures shall be calculated to a datum at the mid-point of the producing section open to flow. The point of entry into the tubing may be used as the datum if it is not more than 100 feet above or below the mid-point of the producing section open to flow.

(2) Under all shut-in conditions and under flowing conditions, when the static column wellhead pressures can be obtained, the bottom-hole pressure shall be calculated as shown in test examples 1 and 2.

(3) When only the flowing wellhead pressure can be obtained, the bottom-hole pressure shall be calculated as shown in test example 3.

(4) When the bottom-hole pressures are recorded by use of a properly calibrated bottom-hole pressure bomb, and corrected to the proper datum, these pressures may be used in the bottom-hole formula.

(5) When liquid accumulation in the well bore during the shut-in period appreciably affects the wellhead shut-in pressure, the calculation of the bottom-hole pressure shall be made as shown in test example 8.

 c. Wellhead Calculations

(1) The static column wellhead pressure must be obtained if possible.

(2) When only the flowing wellhead pressures can be obtained, the static column wellhead pressure shall be calculated as shown in test example 3.

(3) When liquid accumulation in the well bore during the shut-in period appreciably affects the wellhead shut-in pressure, appropriate correction of the surface pressure shall be made. This correction shall be made in the manner shown in test example 8 or, at the option of the operator, by using a bottom-hole pressure bomb and correcting to wellhead conditions as shown in test example 9 or 10.

 4. Reports

Upon completion of the test, all calculations shall be shown on Form BPT 6 and, if applicable, Form BPT 2, BPT 3, or BPT 4. ( ) copies of these forms and the back-pressure curve described below shall be submitted to the (*State Regulatory Body*).

 5. Plotting

A back-pressure curve shall be accurately and neatly drawn on equal log-log paper (3-inch cycles are recommended). This curve shall be drawn through the point established by the stabilized one-point test data and parallel to the most recent curve that has been established for the well by a *Stabilized Multipoint Back-Pressure Test*, a *Constant Time Multipoint Test*, or an *Isochronal Multipoint Test*. The absolute open flow may be determined from this back-pressure curve or calculated as shown in test example 4.

B. DETERMINATION OF DELIVERABILITY

1. Shut-in Pressure

a. Wells with a pipeline connection shall be produced for a sufficient length of time at a flow rate large enough to clear the well bore of accumulated liquids prior to the shut-in period. If the well bore cannot be cleared of accumulated liquids while producing into a pipeline, the well shall be blown to the atmosphere to remove these liquids.

b. Wells without pipeline connections shall be blown to the atmosphere to remove accumulated liquids.

c. The well shall be shut in until the rate of pressure buildup is less than 1/10 of 1 per cent of the previously recorded pressure, psig, in 30 minutes. This pressure shall be recorded.

2. Flow Test

a. After recording the shut-in pressure, a stabilized flow rate and its corresponding pressures shall be taken.

b. This flow rate shall be a rate sufficient to keep the well clear of all liquids.

c. The static column wellhead pressure shall be maintained as nearly as possible at the designated deliverability pressure and within specified tolerances. Any deviation from the specified tolerances must be specifically approved by the (*State Regulatory Body*). Such deviation shall be noted on Form BPT 7.

d. The flow rate measurement shall be obtained by the use of an orifice meter, critical flow prover, positive choke, or other authorized metering device in good operating condition. When an orifice meter is used as the metering device, the meter shall be calibrated and the diameters of the orifice plate and meter run verified as to size, condition, and compliance with acceptable standards. The differential pen shall be zeroed before beginning the test.

e. The field barometric pressure shall be determined.

f. The specific gravity of the separator gas and of the produced liquid shall be determined.

g. Periodically during the flow rate, wellhead flowing pressures and flow rate data shall be recorded and liquid production rate shall be observed to aid in determining when stabilization has been attained. A constant flowing wellhead pressure or static column well-head pressure and rate of flow for a period of at least 15 minutes shall constitute stabilization.

h. At the end of the flow rate the following information shall be recorded: (1) Flow wellhead pressure. (2) Static column wellhead pressure if it can be obtained. (3) Rate of liquid production. (4) Flowing wellhead temperature. (5) All data pertinent to the gas metering device.

3. Calculations

a. The deliverability shall be determined at the designated deliverability pressure by the use of the following formula:

$$D = Q \left[\frac{P_c^2 - P_d^2}{P_c^2 - P_w^2} \right]^n \qquad \text{III–3}$$

b. The static column wellhead pressure shall be obtained if possible.

c. When only the flowing wellhead pressure can be obtained, the static column wellhead pressure shall be calculated as shown in test example 3.

d. When liquid accumulation in the well bore during the shut-in period appreciably affects the wellhead shut-in pressure, appropriate correction of the surface pressure shall be made. This correction shall be made in the manner shown in test example 8 or, at the option of the operator, by using a bottom-hole pressure bomb and correcting to wellhead conditions as shown in text example 9 or 10.

4. Reports

Upon completion of the test, all calculations shall be shown on Form BPT 7 and, if applicable, Form BPT 2, BPT 3, or BPT 4. ( ) copies of these forms shall be submitted to the (*State Regulatory Body*).

IV.
RULES OF CALCULATION

The tables are limited to four significant figures. Significant figures refer to all integers in a number, plus zeros, except those zeros that are in a number for the purpose of indicating the position of the decimal point. Therefore, calculated values shall be rounded off to four significant figures, using the procedure of increasing the last significant figure by one, if followed by a digit of 5, or larger. If the last significant figure is followed by a digit less than 5, it shall remain the same. In calculations involving two or more multipliers, intermediate products should be rounded to five significant figures with the final product being rounded to four significant figures (see table 4–1).

TABLE 4–1 Rounding-off Procedure

Significant Figures			
Five	Four	Three	Two
1.4567	1.457	1.46	1.5
1.4321	1.432	1.43	1.4
0.093451	0.09345	0.0935	0.093*

* Based on 4 significant figures.

In reporting open-flow and deliverability figures, use only whole numbers, limited to four significant figures, for example 590 Mcfd and 63,750 Mcfd. Squared pressures, and the differences in squared pressures, shall be expressed in thousands, limited to the nearest tenth; for example, 1,033.1 and 99.9.

Values of n shall be calculated to four significant figures, and rounded off to three; for example, the calculated value of 0.7583 should be reported as 0.758.

Values of GH/TZ shall be calculated to the fourth figure following the decimal point and rounded off to the third figure following the decimal point.

Values of P_r and T_r shall be calculated to four significant figures and rounded off to two significant figures following the decimal point before determining the Z value. For example, a P_r of 0.3556 would be rounded off to 0.36, a T_r of 1.423 would be rounded off to 1.42.

Exception—High Pressure Wells—An exception from the rules of calculation and significant figures set out above shall be made when making calculations for high pressure wells following procedures outlined in examples 6 and 7 (see complete *IOCC Manual*). In order to provide the necessary accuracy, use 5 or 6 significant figures as indicated by examples 6 and 7.

4.2
State of Texas
Regulations*

RULES OF PROCEDURE

Recent papers on back-pressure testing of Natural Gas Wells have pointed out that the following rules should be observed in order to arrive at a good potential test:

1. The well-bore must be cleaned.

2. The well should be shut in for a period of at least 24 hours, in order to stabilize the reservoir pressure in the vicinity of the well.

3. All pressures should be measured with a dead-weight gauge.

4. The well should be produced at high back-pressure and the rate of flow determined at a minimum of four different working well-head pressures.

When the rate of flow is plotted versus the corresponding value of the difference in the square of the shut-in pressure in the formation and the square of the working pressure at the sand face on logarithmic coordinate paper, the points delineate a straight line which is expressed mathematically by the formula:

$$Q = C(P_f^2 - P_s^2)^n \tag{1}$$

Where:

Q = Rate of flow, Mcf per 24 hours.

C = A numerical coefficient, characteristic of the particular well.

P_f = Shut-in formation pressure, psia.

P_s = Working pressure at the sand face, psia.

n = Numerical exponent, characteristic of the particular well. The value of n may be determined from the slope of the back-pressure curve, plotted in the conventional manner, and obviously is equal to the reciprocal of the slope.

---

*Railroad Commission of Texas, Oil & Gas Division, Back-Pressure Test For Natural Gas Wells State of Texas, P.O. Drawer 12967, Austin, TX 78711.

The test procedure for establishing a back-pressure curve on a natural gas well by the use of a critical flow prover is outlined briefly below:

1. Take the stabilized dead-weight shut-in surface pressure on the well-head.
2. Install the critical flow prover on the well-head, in a vertical position, if possible.
3. Select the correctly sized plate to give the desired pressure drop for the first observation and install it in the prover.
4. Place the thermometer in the well of the prover.
5. Open the valve on the well to allow the full flow of gas to pass into the prover, restricted only by the capacity of the orifice at the operating pressure.
6. The pressure observed on the prover with the dead-weight gauge should be observed and recorded periodically, and the corresponding daily rate of flow should be computed. Whenever two consecutive pressure readings observed over a period of at least 15 minutes agree within 0.1 psig, constant conditions of flow exist and the well is "stabilized."
7. Observe and record the stabilized temperature and pressure readings on the prover at the well head.
8. Shut the well in, install a larger orifice plate in the prover, and again allow the full flow of gas to pass into the prover, and repeat steps 5, 6, 7 and 8.
9. Continue this procedure until at least four sets of stabilized readings have been obtained.

The following general rules have been accepted by the Commission as the approved method for testing natural gas wells by the back-pressure method in the State of Texas. These rules should be followed in order for tests to be accepted by the Engineering Department of the Commission and to eliminate extra work on the part of both the operators and the Commission necessitated by retesting and recalculating results when these rules are not observed.

1. If the well has a pipeline connection, it should be produced at its average daily rate of flow for 24 hours prior to the shut-in period in which the build-up pressure is to be obtained.
2. The well should be shut in for a sufficient length of time to allow a build-up to maximum pressure. The maximum pressure may be

considered attained when the rate of pressure build-up does not exceed one (1) pound per thirty (30) minute period.

3. All pressure readings whether shut-in or flowing should be taken with a deadweight gauge because spring gauge readings are not accurate enough for back-pressure tests.

4. The well should be produced at a rate which is great enough to lift whatever liquid (fluid) may be in the well bore.

5. The test can be run by beginning at either the lowest rate of flow to be employed and proceeding successively to the highest rate, or at the highest rate of flow, and proceeding to the lowest rate. The manner must be shown on the test report.

6. If possible, lower the well-head flowing pressure at least 25% below the shut-in well-head pressure. Some times it may be necessary to produce only one well at a time into a pipeline to achieve this much reduction in pressure.

7. Check the diameter of the orifice plate in the meter-run and also check the inside diameter of the run.

8. The differential pen on the meter should be zeroed.

9. Take pressure readings on the well-head every 15 minutes in order to determine if the well has stabilized.

10. Where a gas well is producing liquid, the gas-liquid ratio should be arrived at from time to time to determine whether or not this ratio remains constant.

11. Under flow conditions, the pressures will be considered stabilized when they do not vary more than 0.1% of the original shut-in well-head pressure during a 15 minute interval.

12. At least 4 rates of flow and 4 corresponding stabilized pressures shall be taken on each test in order that a back-pressure curve may be drawn through at least 3 points.

13. Correct values for compressibility and friction factors should be used in determining the absolute open flow of gas wells.

14(a). A back-pressure curve with a slope of less than 1.0 will not be accepted by the Commission. Specifically this means that when the back-pressure curve is plotted in the conventional manner, the straight line drawn must be at an angle equal to or greater than 45° with the horizontal. Obviously for such conditions, n is equal to or less than 1.0.

14(b). Upon retesting, if the points are not aligned to give a good curve, then a line with a slope of 1.0 should be drawn through the point determined by the highest rate of flow, if the general angle of the points is less than 45° with horizon, or a line with a slope of 0.500 if the general angle of the points is greater than 63.5° with the horizon, and the absolute open flow potential of the gas well should be ascertained from this curve.

15. All necessary data required for calculations of the test should be available in order that these calculations may be made in the field as the test progresses. This procedure will eliminate extra work in the event misleading or incorrect data are obtained.

16. Upon completion of the test, all calculations should be shown on Railroad Commission Form GWT–1 and shall be accompanied by a back-pressure curve neatly plotted on log-log graph paper. Two copies of data sheets, together with back-pressure curves should be mailed to the Commission District Office in which the gas well is located.

4.3
State of
Oklahoma
Regulations*

I. Rule 402†.
Method of Taking Gas Well
Potentials

a. Time and Manner of Taking Potential Tests

Each operator or his agent shall make tests on all gas wells at least once every twelve (12) months for the purpose of determining their potentials, except that such tests shall be made at more frequent intervals by, or upon, direction of the Director of Conservation.

All wells completed in any pool or common source of supply not subject to special pool regulations, shall be tested by the one-point back-pressure method. An assured flow characteristic of 0.85 (exponent n or "slope" of the back-pressure equation) shall be used in establishing the absolute open-flow.

*Oklahoma Corporation Commission, Manual of Back-Pressure Testing of Gas Wells, Thorpe Building, Oklahoma City, OK 73105.
†Quoted from the General Rules and Regulations, Oklahoma Corporation Commission.

Provided further, that at the request of the operator or operators of a majority of the wells in any common source of supply not subject to special pool regulation, the Director of Conservation may direct that such wells be tested by the multi-point back-pressure method or any standard method recognized as being practicable, except that all wells in any one pool or common source of supply shall be tested by the same method.

All official potential tests shall be made in accordance with detailed instructions (such instructions contained herein and including references to the Interstate Oil Compact Commission *Manual of Back-Pressure Testing of Gas Wells*) obtainable from the Conservation Department and shall be witnessed by a field supervisor, engineer, or other party designated by the Conservation Department, and may be observed by an offset operator.

b. Wells Exempt from Potential Tests

Wells having a potential of 1,500,000 cubic feet per day or less, in pools not controlled by special allocation orders, shall be exempt from the potential tests required in Section (a); provided however, that shut-in pressures shall be taken (shut-in pressures shall be taken annually) on all such wells by the operator or his agent and reported on forms prescribed by the Commission.

c. Report of Potential Tests

Reports of potential tests as required by these rules shall be sworn to and forwarded to the Commission, on forms furnished by the Commission for that purpose, at the time they are taken. These reports shall be filed with the Commission immediately upon completion of the tests and shall be signed by the Conservation Department representative who witnessed the tests.

d. Conditioning of Wells

Nothing contained in these rules shall be construed as prohibiting the blowing of wells, when such blowing is necessary for efficient operation.

e. Sections 471 through 477, inclusive, of Title 52 Oklahoma Statutes Annotated, cited as the "Standard Gas Measurement Law," are hereby adopted as rules of the Commission as fully as if set out verbatim herein.

II.
General Instructions

All back-pressure tests required by the Oklahoma Corporation Commission, except in those fields where special field rules apply, shall be conducted in accordance with the procedures set out below.

Gas produced from wells connected to a gas transmission facility shall not be vented to the atmosphere during testing unless absolutely necessary. When an accurate test can be obtained only under conditions requiring venting, the volume vented shall be the minimum required to obtain an accurate test.

All flow measurements shall be obtained by the use of an orifice meter or critical flow prover in good operating condition. When an orifice meter is used as the metering device, the meter shall be calibrated and the diameter of the orifice plate and meter run verified as to size, condition, and compliance with acceptable standards. The differential pen shall be zeroed under operating pressure before beginning the test.

The specific gravity of the separator gas, the produced liquid, and the gas liquid hydrocarbon ratio shall be determined.

The temperature of the gas column must be accurately known to obtain correct test results; therefore, a thermometer well shall be installed in the wellhead. Under shut-in or low flow rate conditions, the observed wellhead temperatures may be distorted by the external temperature. Whenever a thermometer well is not available or when the wellhead temperature has been obviously distorted by the external temperature, a temperature of 60°F shall be used.

Calculations shall be made in the manner prescribed in the appropriate test example. All constants and factors utilized in the calculations shall be obtained from the Interstate Oil Compact Commission *Manual of Back-Pressure Testing of Gas Wells.*

For increased accuracy, the stepwise procedure for computing static column pressures shall be used for all wells having a wellhead shut-in pressure of 2000 psig or greater.

All tests and calculations shall be subject to the review and approval of the Oklahoma Corporation Commission.

All surface pressure readings shall be taken with a dead-weight gauge. Pressure readings taken with a spring gauge will not be accepted.

A. Shut-In Pressures

1. Wells with a pipeline connection shall be produced for at least 24 hours prior to the shut in at a flow rate large enough to clear the well bore of accumulated liquids. If the well bore cannot be cleared of accumulated liquids while producing into a pipeline, the well shall be blown to the atmosphere to remove these liquids.

2. Wells without pipeline connections shall be blown to the atmosphere to remove the accumulated liquids.

3. The shut-in pressure and duration of shut in shall be recorded at any time after the well has been shut in for a minimum of 24 hours.

4. When multiple-completion wells are being tested, all zones shall be shut in at the same time for the purpose of obtaining the shut-in pressure on the zone that is to be tested. This procedure will eliminate any effect a flowing column of gas may have on a static column of gas due to temperature differentials which may exist between the gas columns. The recording of pressures on all zones while shut in and during flow will indicate whether or not communication exists.

5. In the event liquid accumulation in the well bore during the shut-in period appreciably affects the surface pressure, a correction of the indicated surface pressure shall be made by calculating the surface pressure from an accurately determined subsurface pressure. Refer to Test Example 3, Test Example 4, Test Example 5, whichever is applicable.

III.
One-Point Stabilized
Back-Pressure Test Procedure

A. Flow Test

1. After a minimum flow period of six hours, the wellhead flowing pressure and flow rate data shall be recorded at any time stabilization has been reached. The well shall be considered stabilized when the decrease in wellhead flowing pressure is less than 0.1 percent of the previously observed wellhead flowing pressure, psig, during any 15-minute period. If stabilization does not occur within a 24-hour period, data at the end of 24 hours may be utilized.

2. The static column wellhead pressure shall be no more than 90 percent nor less than 75 percent of the wellhead shut-in pressure, psig. If data cannot be obtained in accordance with the foregoing provision, an explanation shall be furnished to the Oklahoma Corporation Commission.

3. At the end of the flow period, the flowing information shall be recorded:

    (a) Flowing wellhead pressure.
    (b) Static column wellhead pressure (if it can be obtained).
    (c) Amount of liquid production.
    (d) Flowing wellhead temperature.
    (e) Duration of flow.

(f) All data pertinent to the gas metering device.
  (1) Prover or line size and orifice size.
  (2) Meter or prover pressure.
  (3) Differential.
  (4) Temperature at point of measurement.
  (5) Type and size of meter or prover.

4. The rate at which the well is producing at the end of the flow period shall be considered the stabilized producing rate corresponding to the static column wellhead pressure existing at that time, provided such rate is not greater than the average producing rate for the entire flow period.

B. Wellhead calculations

1. The wellhead open-flow potential (WHOF) will be determined from the equation:

$$\text{WHOF} = Q \left[\frac{P_c^2}{P_c^2 - p_w^2} \right]^n$$

2. The exponent n will be the n value obtained from the most recent multi-point test which has been approved by the Oklahoma Corporation Commission or the n value assigned to the well by the Commission. (See Section I, General Rule 402(a).)

3. The static column wellhead pressure must be obtained, if possible.

4. When a well has been completed in such a manner that the static column wellhead pressure cannot be obtained, it shall be calculated as shown in Test Example 1 through Test Example 4, as applicable.

5. The average barometric pressure shall be assumed to be 14.40 psia.

IV.
Multi-Point Back-Pressure Test Procedures

A multi-point back-pressure test shall be taken for the purpose of determining the wellhead open-flow potential and exponent n.

A. Flow Tests

1. After recording the shut-in pressure, a series of at least four flow rates of the same duration and the pressures corresponding to each flow rate shall be taken. Each flow shall extend for a maximum period of two hours. If the decrease in wellhead flowing pressure is less than

0.1 percent of the previously observed wellhead flowing pressure, psig, during any 15-minute period prior to the end of the first two-hour flow period, the pressure may be recorded and the next flow started. All subsequent flow periods shall be of the same duration as the first flow period.

2. All rates shall be run in the increasing flow rate sequence. In the case of high liquid ratio wells, or unusual temperature conditions, a decreasing flow rate sequence may be used if the increasing sequence method did not result in the alignment of points. If the decreasing sequence method is used, a statement giving the reason why the use of such method was necessary, together with a copy of the data taken by the increasing sequence method, shall be furnished the Oklahoma Corporation Commission.

3. The lowest flow rate shall be sufficient to keep the well bore clear of all liquids.

4. In order to obtain a good alignment of points, the static column wellhead pressure, psig, at the lowest flow rate should be equal to or less than 95 percent of the shut-in pressure, psig, and at the highest flow rate equal to or greater than 75 percent of the shut-in pressure, psig.

One criterion as to the acceptability of the test is a good spread of data points within the above limits. If data cannot be obtained in accordance with the foregoing provisions, an explanation shall be furnished the Oklahoma Corporation Commission.

5. At the end of each flow rate, the following information shall be recorded:
    (a) Flowing wellhead pressure.
    (b) Static column wellhead pressure (if it can be obtained).
    (c) Amount of liquid production.
    (d) Flowing wellhead temperature.
    (e) Duration of flow.
    (f) All data pertinent to the gas metering device.
        (1) Prover or line size and orifice size.
        (2) Meter or prover pressure.
        (3) Differential.
        (4) Temperature at point of measurement.
        (5) Type and size of meter or prover.

6. The stabilized one-point test data may be obtained by continuation of the last flow rate in the manner prescribed for Flow Test in the *One-Point Stabilization Back-Pressure Test Procedure.* (See Section III, IOCC Manual.)

B. Wellhead Calculations

1. The static column wellhead pressure must be obtained, if possible, at the end of each flow rate.

2. When a well has been completed in such a manner that the static wellhead pressure cannot be obtained, it shall be calculated as shown in Test Example 1 through Test Example 4, as applicable.

3. The average barometric pressure shall be assumed to be 14.40 psia.

C. Plotting

1. The points for the back-pressure curve shall be accurately and neatly plotted on equal scale log-log paper of a minimum of three inches per cycle and a straight line drawn through the best average of three or more points. When no reasonable relationship can be established between three or more points, the well shall be retested.

2. The cotangent of the angle this line makes with the volume coordinate is the exponent n which is used in the back-pressure equation

$$Q = C(P_c^2 - P_w^2)^n$$

The exponent n shall be calculated as shown in Section V, Basic Calculations, Number 5 (IOCC Manual).

3. If the exponent n is greater than 1.000 or less than 0.500, the well shall be retested.

4. If, after retesting a well, a satisfactory test is not obtained, the Oklahoma Corporation Commission may grant an exception and assign a value of the exponent n to the well.

D. Calculation of Wellhead Open-Flow Potential

Using the pressure and volume corresponding with the highest rate of flow which falls on the curve, calculate the wellhead open flow potential from the equation

$$\text{WHOF} = Q \left[\frac{P_c^2}{P_c^2 - P_w^2} \right]^n$$

V.
Reporting

A. All shut-in pressures required by General Rule 402(b) shall be reported on Commission Form 1005.

B. One-point stabilized back-pressure tests shall be reported on Commission Form 1016.

C. Multi-point back-pressure tests shall be reported on Commission Form 1016 with back-pressure curve attached.

D. Field data sheet for both one-point and multi-point back-pressure tests shall be reported on Commission Form 1016a.

E. The following Commission forms will also be required:

    1. Form 1016b, Worksheet for Calculations of Static Column Wellhead Pressure (P_w), when using average temperature and compressibility factors.

    2. Form 1016c, Worksheet for Calculation of Static Column Wellhead Pressure (P_w), when using the stepwise method.

    3. Form 1016d, Worksheet for Calculation of Static Column Wellhead Pressure (P_c or P_w), when the observed wellhead pressure is affected by liquids in the wellbore.

4.4
State of
New Mexico
Regulations*

I.
General Instructions

All back-pressure tests required by the New Mexico Oil Conservation Commission shall be conducted in accordance with the procedures set out below except for those wells in pools where special testing procedures are applicable.

> Note: Order R-333-F, and any amendments thereto, prescribes Procedures for testing wells in a number of the fields in the San Juan Basin Area of San Juan, Rio Arriba, and Sandoval Counties, New Mexico.

In general, wells with pressures less than 2,000 psig shut-in wellhead pressure may be calculated by the simplified method using average

*State of New Mexico Oil Conservation Commission, *Manual for Back-Pressure Testing of Natural Gas Wells*, Order R-333-F, January 1, 1966, P.O. Box 2088, Santa Fe, NM 87501.

temperature and super-compressibility as shown in Test Example 1A. Wells with pressures in excess of 2,000 psig shut-in wellhead pressures should be calculated using the stepwise procedure as shown in Test Example 1B. Calculations shall be made in the manner prescribed in the appropriate test examples. The observed data and calculations shall be reported on the prescribed forms.

Gas produced from wells connected to a gas transportation facility should not be vented to the atmosphere during testing. When an accurate test can be obtained only under conditions requiring venting, the volume vented shall be the minimum required to obtain an accurate test.

Note: Flow periods for unconnected wells shall be one hour unless stabilization is reached in a lesser time.

All surface pressure readings shall be taken with a dead weight gauge. Subsurface pressures determined by the use of a properly calibrated bottom hole pressure gauge are acceptable.

The temperature of the gas column must be accurately known to obtain correct test results; therefore, a thermometer well should be installed in the wellhead. Under shut-in or low flow rate conditions, the observed wellhead temperatures may be distorted by the external temperature. Whenever this situation exists, the mean annual temperature should be used. This shall be 50°F for Northwest New Mexico and 60°F for Southeast New Mexico.

When necessary to make friction calculations for the flowing column, the length "*L*" (length of the flow channel in feet) shall be measured from surface to the top of the producing section open to flow. The point of entry into the tubing may be used as "*L*" if it is not more than 100 ft. above or below the top of the producing section open to flow. If the bottom of the tubing is more than 100 ft. above or below the top of the producing section, the casing diameter and the length from this point to the bottom of the tubing should be considered in making friction calculations. For uncased hole below the base of the tubing or casing show, friction calculations may be ignored.

II.
Multipoint Back-Pressure Test Procedures

A multipoint back-pressure test shall be taken for the purpose of determining the absolute open flow and exponent *n* from the plot of the wellhead equation:

$$Q = C(P_c^2 - P_w^2)^n$$

A. Stabilized Multipoint Test

1. Shut-in Pressure

a. Wells with a pipeline connection shall be produced for a sufficient length of time at a flow rate large enough to clear the well bore of accumulated liquids prior to the shut-in period. If the well bore cannot be cleared of accumulated liquids while producing into a pipeline, the well shall be blown to the atmosphere to remove these liquids.

b. Wells without pipeline connections shall be blown to the atmosphere to remove the accumulated liquids.

c. The well shall be shut in until the rate of pressure buildup is less than 1/10 of 1 per cent of the previously recorded pressure, psig, in 30 minutes. This pressure shall be recorded.

2. Flow Tests

a. After recording the shut-in pressure, a series of at least four stabilized flow rates and the pressure corresponding to each flow rate shall be taken (see paragraph g). Any shut-in time between flow rates shall be held to a minimum. These rates shall be run in the increasing flow-rate sequence. In the case of high liquid ratio wells or unusual temperature conditions, a decreasing flow-rate sequence may be used if the increasing sequence method did not result in point alignment. If the decreasing sequence method is used, a statement giving the reasons why the use of such method was necessary, together with a copy of the data taken by the increasing sequence method, shall be furnished to the New Mexico Oil Conservation Commission. If experience has shown that the use of the decreasing sequence method is necessary for an accurate test, a test by the increasing sequence method will not be required.

b. The lowest flow rate shall be a rate sufficient to keep the well clear of all liquids.

c. One criterion as to the acceptability of the test is a good spread of data points. In order to assure a good spread of points, the flowing wellhead pressure, psig, at the lowest flow rate should not be more than 95 per cent of the well's shut-in pressure, psig, and at the highest flow rate not more than 75 per cent of the well's shut-in pressure, psig. If data cannot be obtained in accordance with the foregoing provisions, an explanation shall be furnished to the New Mexico Oil Conservation Commission.

d. All flow-rate measurements shall be obtained by the use of an orifice meter, critical flow prover, positive choke, or other authorized metering device in good operating condition. When an orifice meter is used as the metering device, the meter shall be calibrated and the diameters of the orifice plate and meter run verified as to size, condition, and compliance with acceptable standards. The differential pen shall be zeroed under operating pressure before beginning the test.

e. The field barometric pressure shall be 12.2 psia for the Northwest and 13.2 psia for the Southeast.

f. The specific gravity of the separator gas and of the produced liquid shall be determined.

g. Periodically, during each flow rate, flowing wellhead pressures and flow-rate data shall be recorded and liquid production rate shall be observed to aid in determining when stabilization has been attained. A constant flowing wellhead pressure or static column wellhead pressure and rate of flow for a period of at least 15 minutes shall constitute stabilization for this test. *For unconnected wells the flow periods for this test are limited to one hour for each rate of flow.* Longer flow periods of unconnected wells shall not be made without special permission from the Commission. If stabilization is not obtained, the procedure for conducting a Non-Stabilized Multipoint Test shall be used.

h. At the end of each flow rate the following information shall be recorded: (1) Flowing wellhead pressure. (2) Static column wellhead pressure if it can be obtained. (3) Rate of liquid production. (4) Flowing wellhead temperature. (5) All data pertinent to the gas metering device.

3. Calculations

a. General

The absolute open flow is determined from the wellhead equation $Q = C(P_c^2 - P_w^2)^n$.

b. Wellhead Calculations

(1) The static column wellhead pressure (P_w) must be obtained if possible.

(2) When only the flowing wellhead pressure (P_t) can be obtained, the static column wellhead pressure shall be calculated as shown in Test Example 1.

(3) When liquid accumulation in the well bore during the shut-in period appreciably affects the shut-in wellhead pressure, appropriate correction of the surface pressure shall be made. This correction shall be made in the manner shown in Test Example 7 or, at the option of the operator, by using a bottom-hole pressure gauge and correcting to wellhead conditions as shown in Test Example 5 or 6.

4. Reports

Upon completion of the test, all calculations shall be shown on Form C-122. Two copies of these forms and the back-pressure curve described below shall be submitted to the appropriate district office of the

New Mexico Oil Conservation Commission along with one copy of the calculations on the appropriate form as indicated in the Test Examples.

5. Plotting

a. The points (Q vs. $P_c^2 - P_w^2$) for the back-pressure curve shall be accurately and neatly plotted on equal scale log-log paper (3-inch cycles are recommended) and a straight line drawn through the best average of three or more points. When no reasonable relationship can be established between three or more points, the well shall be retested.

b. The cotangent of the angle this line makes with the volume coordinate is the exponent n which is used in the back-pressure equation. The exponent n shall always be calculated as shown in Basic Calculation 5.

c. If the exponent n is greater than 1.000 or less than 0.500, the well shall be retested.

d. If, after retesting the well, no reasonable alignment is established between three or more points, then a straight line shall be drawn through the best average of at least three points *of the retest.*

(1) If the exponent n is greater than 1.000, a straight line with an exponent n of 1.000 shall be drawn through the point corresponding to the highest flow rate utilized in establishing the line.

(2) If the exponent n is less than 0.500, a straight line with an exponent n of 0.500 shall be drawn through the point corresponding to the lowest flow rate utilized in establishing the line.

B. Non-Stabilized Multipoint Tests

When well stabilization is impractical to obtain for a series of points, the exponent n of the back-pressure curve shall be established by either the Constant Time Multipoint Test or the Isochronal Multipoint Test. The exponent n so determined shall then be applied to a stabilized one-point test to determine the absolute open flow. (See *Stabilized One-Point Back-Pressure Test Procedure*, pg. 12.) The flow during this one-point test shall be for a period adequate to reach stabilized conditions unless determination is made in conjunction with the New Mexico Oil Conservation Commission that it would be impractical to continue flow until complete stabilization is reached. The one-point stabilized flow shall not be taken on unconnected wells and the absolute open flow shall be calculated on the basis of four one-hour flows.

1. Constant Time Multipoint Test

a. Shut-in Pressure

(1) Wells with a pipeline connection shall be produced for a sufficient length of time at a flow rate large enough to clear the well bore of accumulated liquids prior to the shut-in period. If the well bore cannot be cleared of accumulated liquids while producing into a pipe-

line, the well shall be blown to the atmosphere to remove these liquids.

(2) Wells without pipeline connections shall be blown to the atmosphere to remove accumulated liquids.

(3) The well shall be shut in until the rate of pressure buildup is less than 1/10 of 1 per cent of the previously recorded pressure, psig, in 30 minutes. This pressure shall be recorded.

b. Flow Tests

(1) After recording the shut-in pressure, a series of at least four flow rates of the same duration and the pressures corresponding to each flow rate shall be taken. *For unconnected wells, each flow rate shall be one hour.* Any shut-in time between flow rates shall be held to a minimum. These rates shall be run in the increasing flow-rate sequence. In the case of high liquid ratio wells or unusual temperature conditions, a decreasing flow-rate sequence may be used if the increasing sequence method did not result in the alignment of points. If the decreasing sequence method is used, a statement giving the reasons why the use of such method was necessary, together with a copy of the data taken by the increasing sequence method, shall be furnished the New Mexico Oil Conservation Commission. If experience has shown that the use of the decreasing sequence method is necessary for an accurate test, a test by the increasing sequence method will not be required.

(2) The lowest flow rate shall be a rate sufficient to keep the well clear of all liquids.

(3) One criterion as to the acceptability of the test is a good spread of data points. In order to assure a good spread of points, the flowing wellhead pressure, psig, at the lowest flow rate should not be more than 95 per cent of the well's shut-in pressure, psig, and at the highest flow rate not more than 75 per cent of the well's shut-in pressure, psig. If data cannot be obtained in accordance with the foregoing provisions, an explanation shall be furnished to the New Mexico Oil Conservation Commission.

(4) All flow-rate measurements shall be obtained by the use of an orifice meter, critical flow prover, positive choke, or other authorized metering device in good operating condition. When an orifice meter is used as the metering device, the meter shall be calibrated and the diameters of the orifice plate and meter run verified as to size, condition, and compliance with acceptable standards. The differential pen shall be zeroed under operating pressure before beginning the test.

(5) The field barometric pressure shall be 12.2 psia for the Northwest and 13.2 psia for the Southeast.

(6) The specific gravity of the separator gas and of the produced liquid shall be determined.

(7) At the end of each flow rate the following information shall be recorded: (a) Flowing wellhead pressure. (b) Static column wellhead pressure if it can be obtained. (c) Rate of liquid production. (d) Flowing wellhead temperature. (e) All data pertinent to the gas metering device.

(8) The stabilized one-point test data may be obtained by continuation of the last flow rate in the manner prescribed for Flow Test in the *Stabilized One-Point Back-Pressure Test Procedure*. The one-point stabilized flow shall not be taken on unconnected wells and the absolute open flow shall be determined on the basis of four one-hour flows.

c. Calculations

(1) General

The absolute open flow is determined from the wellhead equation $Q = C(P_c^2 - P_w^2)^n$.

(2) Wellhead Calculations

(a) The static column wellhead pressure (P_w) must be obtained if possible.

(b) When only the flowing wellhead pressure (P_t) can be obtained, the static column wellhead pressure shall be calculated as shown in Test Example 1.

(c) When liquid accumulation in the well bore during the shut-in period appreciably affects the shut-in wellhead pressure, appropriate correction of the surface pressure shall be made. This correction shall be made in the manner shown in Test Example 7 or, at the option of the operator, by using a bottom-hole pressure gauge and correcting to wellhead conditions as shown in Test Example 5 or 6.

d. Reports

Upon completion of the test, all calculations shall be shown on Form C–122. Two copies of these forms and the back-pressure curve described below shall be submitted to the appropriate district office of the New Mexico Oil Conservation Commission along with one copy of the calculations on the appropriate form as indicated in the Test Examples.

e. Plotting

(1) The points (Q vs. $P_c^2 - P_w^2$) for the back-pressure curve shall be accurately and neatly plotted on equal scale log-log paper (3-inch cycles are recommended) and a straight line drawn through the best average of three or more points. When no reasonable relationship can be established between three or more points, the well shall be retested.

(2) The cotangent of the angle this line makes with the volume coordinate is the exponent *n* which is used in the back-pressure equa-

tion. The exponent *n* shall always be calculated as shown in Basic Calculation 5.

(3) If the exponent *n* is greater than 1.000 or less than 0.500, the well shall be retested.

(4) If, after retesting the well, no reasonable alignment is established between three or more points, then a straight line shall be drawn through the best average of at least three points *of the retest.*

(a) If the exponent *n* is greater than 1.000, a straight line with an exponent *n* of 1.000 shall be drawn through the point corresponding to the highest flow rate utilized in establishing the line.

(b) If the exponent *n* is less than 0.500, a straight line with an exponent *n* of 0.500 shall be drawn through the point corresponding to the lowest rate of flow utilized in establishing the line.

(5) The constant time data points are used only to determine the value of the exponent *n*. The back-pressure curve shall be drawn through the stabilized data point and parallel to the line established by the constant time data points. For unconnected wells, the back-pressure curve shall be drawn through the points representing the four one-hour flow rates. The absolute open flow may be determined from this back-pressure curve or calculated as shown in Test Example 3.

2. Isochronal Multipoint Test

a. Shut-In Pressures

(1) Wells with a pipeline connection shall be produced for a sufficient length of time at a flow rate large enough to clear the well bore of accumulated liquids prior to the shut-in period. If the well bore cannot be cleared of accumulated liquids while producing into a pipeline, the well shall be blown to the atmosphere to remove these liquids.

(2) Wells without pipeline connections shall be blown to the atmosphere to remove accumulated liquids.

(3) Prior to each flow test as described below, the well shall be shut-in until the rate of pressure build-up is less than 1/10 of 1 per cent of the previously recorded pressure, psig, in 30 minutes. This pressure shall be recorded and used with the data from the subsequent flow test.

b. Flow Tests

(1) After recording the initial shut-in pressure, a series of at least four flow rates of the same duration and the pressures corresponding to each flow rate shall be taken. For unconnected wells, each flow rate shall be one hour. Each flow rate shall be preceded by a shut-in pressure as prescribed above in 2.a.(3).

(2) The lowest flow rate shall be a rate sufficient to keep the well clear of all liquids.

(3) One criterion as to the acceptability of the test is a good

spread of data points. In order to assure a good spread of points, the flowing wellhead pressure, psig, at the lowest flow rate should not be more than 95 per cent of the well's shut-in pressure, psig, and at the highest flow rate not more than 75 per cent of the well's shut-in pressure, psig. If data cannot be obtained in accordance with the foregoing provisions, an explanation shall be furnished to the New Mexico Oil Conservation Commission.

(4) All flow-rate measurements shall be obtained by the use of an orifice meter, critical flow prover, positive choke, or other authorized metering device in good operating condition. When an orifice meter is used as the metering device, the meter shall be calibrated and the diameters of the orifice plate and meter run verified as to size, condition, and compliance with acceptable standards. The differential pen shall be zeroed under operating pressure before beginning the test.

(5) The field barometric pressure shall be 12.2 psia for the Northwest and 13.2 psia for the Southeast.

(6) The specific gravity of the separator gas and of the produced liquid shall be determined.

(7) At the end of each flow rate the following information shall be recorded: (a) Flowing wellhead pressure. (b) Static column wellhead pressure if it can be obtained. (c) Rate of liquid production. (d) Flowing wellhead temperature. (e) All data pertinent to the gas metering device.

(8) The stabilized one-point test data may be obtained by continuation of the last flow rate in the manner prescribed for Flow Test in the *Stabilized One-Point Back-Pressure Test Procedure.* The one-point stabilized flow shall not be taken on unconnected wells and the absolute open flow shall be determined on the basis of four one-hour flows.

c. Calculations

(1) General

The absolute open flow is determined from the wellhead equation $Q = C(P_c^2 - P_w^2)^n$.

(2) Shut-In Pressure

(a) The shut-in pressure preceding each flow rate shall be used in conjunction with the static column wellhead pressure corresponding to that flow.

(3) Wellhead Calculations

(a) The static column wellhead pressure (P_w) must be obtained if possible.

(b) When only the flowing wellhead pressure (P_t) can be obtained, the static column wellhead pressure shall be calculated as shown in Test Example 1.

(c) When liquid accumulation in the well bore during the shut-in period appreciably affects the shut-in wellhead pressure, appropriate correction of the surface pressure shall be made. This correction shall be made in the manner shown in Test Example 7 or, at the option of the operator, by using a bottom-hole pressure gauge and correcting to wellhead conditions as shown in Test Example 5 or 6.

d. Reports

Upon completion of the test, all calculations shall be shown on Form C–122. Two copies of these forms and the back-pressure curve described below shall be submitted to the appropriate district office of the New Mexico Oil Conservation Commission along with one copy of the calculations on the appropriate form as indicated in the Test Examples.

e. Plotting

(1) The points (Q vs. $P_c^2 - P_w^2$) for the back-pressure curve shall be accurately and neatly plotted on equal scale log-log paper (3-inch cycles are recommended) and a straight line drawn through the best average of three or more points. When no reasonable relationship can be established between three or more points, the well shall be retested.

(2) The cotangent of the angle this line makes with the volume coordinate is the exponent n which is used in the back-pressure equation. The exponent n shall always be calculated as shown in Basic Calculation 5.

(3) If the exponent n is greater than 1.000 or less than 0.500, the well shall be retested.

(4) If, after retesting the well, no reasonable alignment is established between three or more points, then a straight line shall be drawn through the best average of at least three points *of the retest*.

(a) If the exponent n is greater than 1.000, a straight line with an exponent n of 1.000 shall be drawn through the point corresponding to the highest rate of flow utilized in establishing the line.

(b) If the exponent n is less than 0.500, a straight line with an exponent n of 0.500 shall be drawn through the point corresponding to the lowest rate of flow utilized in establishing the line.

(5) The isochronal data points are used only to determine the value of the exponent n. The back-pressure curve shall be drawn through the stabilized data point and parallel to the line established by the Isochronal data points. For unconnected wells, the back-pressure curve shall be drawn through the points representing the four one-hour flow rates. The absolute open flow may be determined from this back-pressure curve or calculated as shown in Test Example 3.

(This shall not be used for unconnected wells) The most recently determined exponent *n* established by a *Stabilized Multi-Point Back-Pressure Test*, a *Constant Time Multipoint Test*, or an *Isochronal Multipoint Test* shall be used with this stabilized one-point test to determine the absolute open flow.

The Flow Test portion of this test may be used in conjunction with the *Constant Time Multipoint Test* or the *Isochronal Multipoint Test* to determine the absolute open flow.

A. Determination Of Absolute Open Flow

  1. Shut-in Pressure

    a. The well shall be produced for a sufficient length of time at a flow rate large enough to clear the well bore of accumulated liquids prior to the shut-in period. If the well bore cannot be cleared of accumulated liquids while producing into a pipeline, the well shall be blown to the atmosphere to remove these liquids.

    b. The well shall be shut in until the rate of pressure build-up is less than 1/10 of 1 per cent of the previously recorded pressure, psig, in 30 minutes. This pressure shall be recorded.

  2. Flow Test

    a. After recording the shut-in pressure, a stabilized flow rate and its corresponding pressures shall be taken.

    b. This flow rate shall be a rate sufficient to keep the well clear of all liquids.

    c. The flowing wellhead pressure, psig, shall not be more than 95 per cent of the well's shut-in pressure, psig. If the data cannot be obtained in accordance with the foregoing provision, an explanation shall be furnished to the New Mexico Oil Conservation Commission.

    d. The flow-rate measurement shall be obtained by the use of an orifice meter, or other authorized metering device in good operating condition. When an orifice meter is used as the metering device, the meter shall be calibrated and the diameters of the orifice plate and meter run verified as to size, condition, and compliance with acceptable standards. The differential pen shall be zeroed under operating pressure before beginning the test.

    e. The field barometric pressure shall be 12.2 psia for the Northwest and 13.2 psia for the Southeast.

    f. The specific gravity of the separator gas and of the produced liquid shall be determined.

    g. Periodically during the flow rate, flowing wellhead pressures and flow-rate data shall be recorded and liquid production rate shall be observed to aid in determining when stabilization has been attained. A constant flowing wellhead pressure or static column wellhead pressure and rate of flow for a period of at least 15 minutes shall constitute stabilization for this test.

h. At the end of the flow rate the following information shall be recorded. (1) Flowing wellhead pressure. (2) Static column wellhead pressure if it can be obtained. (3) Rate of liquid production. (4) Flowing wellhead temperature. (5) All data pertinent to the gas metering device.

3. Calculations

a. General

The absolute open flow is determined from the wellhead equation $Q = C(P_c^2 - P_w^2)^n$.

b. Wellhead Calculations

(1) The static column wellhead pressure (P_w) must be obtained if possible.

(2) When only the flowing wellhead pressure (P_t) can be obtained, the static column wellhead pressure shall be calculated as shown in Test Example 1.

(3) When liquid accumulation in the well bore during the shut-in period appreciably affects the shut-in wellhead pressure, appropriate correction of the surface pressure shall be made. This correction shall be made in the manner shown in Test Example 7 or, at the option of the operator, by using a bottom-hole pressure gauge and correcting to wellhead conditions as shown in Test Example 5 or 6.

4. Reports

Upon completion of the test, all calculations shall be shown on Form C–122. Two copies of this form and the *latest multipoint back-pressure curve* shall be submitted to the appropriate district office of the New Mexico Oil Conservation Commission along with one copy of the calculations on the appropriate form as indicated in the Test Examples.

B. Determination of Deliverability

(This shall not be used for unconnected wells)

1. Shut-in Pressure

a. The well shall be produced for a sufficient length of time at a flow rate large enough to clear the well bore of accumulated liquids prior to the shut-in period. If the well bore cannot be cleared of accumulated liquids while producing into a pipeline, the well shall be blown to the atmosphere to remove these liquids.

b. The well shall be shut in until the rate of pressure build-up is less than 1/10 of 1 per cent of the previously recorded pressure, psig, in 30 minutes. This pressure shall be recorded.

2. Flow Test

a. After recording the shut-in pressure, a stabilized flow rate and its corresponding pressures shall be taken.

b. This flow rate shall be a rate sufficient to keep the well clear of all liquids.

c. The static column wellhead pressure shall be maintained as nearly as possible at the designated deliverability pressure and within specified tolerances. Any deviation from the specified tolerances must be specifically approved by the New Mexico Oil Conservation Commission. Such deviation shall be noted on Form No. C–122–C.

d. The flow-rate measurement shall be obtained by the use of an orifice meter, or other authorized metering device in good operating condition. When an orifice meter is used as the metering device, the meter shall be calibrated and the diameters of the orifice plate and meter run verified as to size, condition, and compliance with acceptable standards. The differential pen shall be zeroed before beginning the test.

e. The field barometric pressure shall be 12.2 psia for the Northwest and 13.2 psia for the Southeast.

f. The specific gravity of the separator gas and of the produced liquid shall be determined.

g. Periodically during the flow rate, flowing wellhead pressures and flow-rate data shall be recorded and liquid production rate shall be observed to aid in determining when stabilization has been attained. A constant flowing wellhead pressure or static column wellhead pressure and rate of flow for a period of at least 15 minutes shall constitute stabilization.

h. At the end of the flow rate the following information shall be recorded. (1) Flowing wellhead pressure. (2) Static column wellhead pressure if it can be obtained. (3) Rate of liquid production. (4) Flowing wellhead temperature. (5) All data pertinent to the gas metering device.

3. Calculations

a. The deliverability shall be determined at the designated deliverability pressure by the use of the following formula:

$$D = Q \left[\frac{P_c^2 - P_d^2}{P_c^2 - P_w^2} \right]^n$$

b. The static column wellhead pressure (P_w) shall be obtained if possible.

c. When only the flowing wellhead pressure (P_f) can be obtained, the static column wellhead pressure shall be calculated as shown in Test Example 1.

d. When liquid accumulation in the well bore during the shut-in period appreciably affects the shut-in wellhead pressure, appropriate correction of the surface pressure shall be made. This correction shall be made in the manner shown in Test Example 7 or, at the option of the operator, by using a bottom-hole pressure gauge and correcting to wellhead conditions as shown in Test Example 5 or 6.

4. Reports

Upon completion of the test, all calculations shall be shown on Form C–122–CO. Two copies of this form and the latest multipoint back-pressure curve shall be submitted to the appropriate district office of the New Mexico Oil Conservation Commission.

IV.
Rules Of Calculation

The tables are limited to four significant figures. Significant figures refer to all integers in a number, plus zeros, except those zeros that are in a number for the purpose of indicating the position of the decimal point. Therefore, calculated values shall be rounded off to four significant figures, using the procedure of increasing the last significant figure by one, if followed by a digit of 5, or larger. If the last significant figure is followed by a digit less than 5, it shall remain the same. In calculations involving two or more multipliers, intermediate products should be rounded to five significant figures with the final product being rounded to four significant figures (see table 4–2).

TABLE 4.2 Rounding-off Procedure

Significant Figures			
Five	Four	Three	Two
1.4567	1.457	1.46	1.5
1.4321	1.432	1.43	1.4
0.093451	0.09345	0.0935	0.093*

*Based on four significant figures.

In reporting absolute open flow and deliverability figures, use only whole numbers, limited to four significant figures; for example 590 Mcfd and 63,750 Mcfd. Squared pressures, and the differences in squared pressures, shall be expressed in thousands, limited to the nearest tenth; for example 1,033.1 and 99.9.

Values of n shall be calculated to four significant figures, and rounded off to three; for example, the calculated value of 0.7583 should be reported as 0.758.

Values of GH/TZ shall be calculated to the fourth figure following the decimal point and rounded off to the third figure following the decimal point.

Values of P_r and T_r shall be calculated to four significant figures and rounded off to two significant figures following the decimal point before determining the Z value. For example, a P_r of 0.3556 would be rounded off to 0.36, a T_r of 1.423 would be rounded off to 1.42.

Exception—High Pressure Wells. An exception from the rules of calculation and significant figures set out above shall be made when making calculations for high pressure wells. In order to provide the necessary accuracy, use five or six significant figures.

4.5 State of Louisiana Regulations*

Rules Of Procedure

Gas wells shall not be tested by the "open-flow" method. The back-pressure method of determining the open-flow, as outlined by the Bureau of Mines in their Monograph 7, *Back-Pressure Data on Natural Gas Wells*, shall be used. When, for any reasons, the back-pressure method is not feasible, an acceptable method, not entailing excessive physical waste of gas, may be used, upon recommendation of the technical staff of the Department.

Louisiana Oil & Gas Handbook, Sixth edition, First Supplement March 1979, Act 157 of 1940 as amended by Statewide Order No. 29–B, Section X: Well Allowables and Completion (H). 1260 Havenwood Drive, Baton Rouge, LA 70815.

4.6 Province of Alberta Regulations*

Rules Of Procedure
11.130

(1) Any gas well test made pursuant to the requirements of these regulations shall be conducted in accordance with procedures described in the Board publication Guide G–3, *Gas Well Testing—Theory and Practice*, Fourth Edition (SI).

(2) Unless the well is completed in a designated pool from which production other than test production has commenced, the licensee of a gas well being tested pursuant to the requirements of these regulations shall obtain a satisfactory subsurface static pressure with a properly calibrated subsurface gauge, but the Board may grant relief from this requirement where special circumstances warrant.

(3) Prior to the first regular production from a gas well in a pool where allowables are prescribed by Board order, the licensee shall test the well by a multi-point deliverability test using a properly calibrated subsurface pressure gauge.

(4) Prior to or during the first 90 days of production from a gas well in a pool where allowables are not prescribed by Board order, the licensee shall test the well by a multi-point deliverability test.

(5) The licensee of any producing gas well shall conduct a single-point deliverability test on the well during the second twelve-month period following the commencement of regular production.

(6) The Board may, upon its own motion or upon application therefor, grant relief from the provisions of subsections (4) and (5) hereof where the well is clearly of an infill type, or where the stabilized deliverability relationship of the well is to be obtained by some other test acceptable to the Board.

(7) Where a gas well is granted relief pursuant to subsection (6)

---

*Oil and Gas Conservation Regulations of the Province of Alberta, Section 11.130. Energy Resources Conservation Board, 603 Sixth Avenue SW, Calgary, Alberta, Canada T2P 0T4.

hereof the licensee shall, prior to or during the first 90 days of regular production from the well, obtain the stabilized shut-in reservoir pressure of the well.

(8) Prior to or during the first 90 days of regular production from a gas well which has received any reservoir stimulation treatment or change in production equipment that will significantly alter its ability to deliver gas, the licensee shall test the well by a multi-point deliverability test.

(9) Where a gas well is tested in accordance with this section, the licensee shall ensure that a properly calibrated subsurface pressure gauge is used whenever liquid accumulation in the wellbore is known or suspected to be present during the conduct of the test.

(10) The licensee or an appointed test co-ordinator of any producing gas well completed in a pool containing published marketable gas reserves greater than 300 million cubic metres shall, prior to April 30th of each year, submit to the Board a schedule of testing which provides for alternate multi-point and single-point deliverability tests every three years throughout the producing life of the well, each test to include the accurate determination of stabilized shut-in reservoir pressure, but the Board may modify these requirements where either

(a) the stabilized deliverability relationship of the well is to be obtained by some other test acceptable to the Board, or

(b) a sufficient number of representative wells spaced throughout the pool are scheduled for appropriate testing.

(11) The results of any gas well test required to be reported to the Board shall be reported by the licensee or an appointed test co-ordinator on either

(a) appropriate forms available from the Board, or

(b) such other forms as the Board may consider appropriate.

(12) Notwithstanding subsections (5) and (8) hereof the Board may, upon its own motion or upon application therefor, grant relief from the provisions of this section, but where the Board grants such relief it may do so subject to such terms, conditions or other requirements as it considers appropriate in the circumstances.

Chapter 5

Field Test & Interpretation

In this chapter we shall undertake the analysis of an actual well test. The well to be tested, the Melancon Heirs #1, is located in Acadia Parish, Louisiana, about 12 miles due west of Lafayette. It is a North Rayne Field stepout well and was perforated and cleaned up just prior to the test. Both gas and liquid production were expected, and, because a pipeline was not available, it was decided to flare the produced hydrocarbons and dispose any water production in the pit.

During the conventional backpressure test, the procedure specified by the State of Louisiana, stabilization was achieved rather fast, and both liquid and gas production were encountered. Because of the presence of liquids, the surface equipment included two stages of separation. There was no indication of coning or sand production. The well was produced in the reverse sequence of rates, that is, from high to low, to prevent liquid buildup in the well bore. There was no H_2S present in the gas, pressures were measured at the wellhead using both a deadweight tester and a continuous recorder and depth measurements are vertical. A photograph of the testing equipment is shown in figure 5.1.

Tables 5.1, 5.2, and 5.3 give the well data, summary of test data, and calculation of the constants for a deliverability test for the Melancon Heirs # 1, respectively.

TABLE 5.1 Well Data

TBG Size and Weight:	2 7/8", 6.50#
CSG Size and Weight:	5 1/2", 20#
P.B.T.D.:	13,593'
Packer Setting:	11,935'
Perforations:	12,758'–12,778'
Tubing I.D.:	2.441

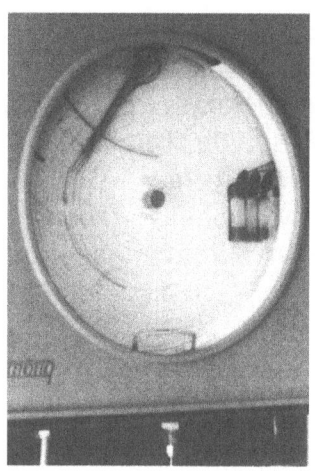

Figure 5.1
Pressure Recorder and Critical Flow Prover Used during Gaswell Tests.

TABLE 5.2 Summary Of Test Data.

	Test No. 1	Test No. 2	Test No. 3	Test No. 4
Length of Test	4 hours	3 hours	3 hours	3 hours
Wellhead Pressures				
Shut-In	4261 psig	4261 psig	4261 psig	4261 psig
Flowing (Initial)	3815 psig	4060 psig	3970 psig	4045 psig
Flowing (Final)	3800 psig	3920 psig	3985 psig	4045 psig
Casing	800 psig	800 psig	80 psig	40 psig
Oil Production Rate	180 STB/D	132 STB/D	103.92 STB/D	48 STB/D
Water Production Rate	12 STB/D	Trace	Trace	Trace
Gas Production Rate	3,921 MSCF/D	3,199 MSCF/D	2,510 MSCF/D	1,910 MSCF/D

TABLE 5.3 Calculation Of The Constants For a Deliverability Test On The Melancon #1, West of Lafayette, LA.

	q_{sc} MMSCF/D	p_{wf} (wellhead) psia	p_{wf} (bottomhole) psia	p_{wf}^2 (psia)^2	$\bar{p}_r^2 - p_{wf}^2$ (psia)^2
Shut-in	0	4276	5260 = $\bar{p}_r$	27.668×10^6	
Flow 1	3.921	3815	4792	25.412×10^6	2.256×10^6
Flow 2	3.199	3935	4923	24.751×10^6	2.917×10^6
Flow 3	2.510	4000	4975	24.236×10^6	3.432×10^6
Flow 4	1.910	4060	5041	22.963×10^6	4.705×10^6

All the calculations are based on a datum depth of 12,768 ft, the midpoint of the perforated interval. The following additional data is required: Gas Gravity = 0.625 (Air = 1.00), Reservoir Temperature = 707°R, Atmospheric Pressure = 15 psia.

The Hurst and Bellis* formula was used to calculate bottomhole pressures from surface pressure.

---

*Hurst and Bellis calculation is a relatively short and practical technique that can be carried out at the wellsite during the test. The following empirical formula of Hurst and Bellis can be solved for the following sandface pressure or the shut-in bottomhole pressure:

$$P_{wf \text{ (bottomhole)}} = P_{wf \text{ (wellhead)}} \sqrt{e^k + F(e^k - 1)}$$

where

$$k = 0.037502 \, \frac{\rho_g L}{Tz}$$

$$\log F = 1.11352 + 1.82930 \log \frac{zQT}{d^{3.24832} P_{wf \text{ (wellhead)}} \, \rho_g}$$

$P_{wf \text{ (wellhead)}}$ = wellhead pressure, psia

$P_{wf \text{ (bottomhole)}}$ = flowing bottomhole pressure, psia

ρ_g = gas gravity, (air = 1)

L = depth to the midpoint of the perforations, feet

T = average flowing temperature in tubing, °R

z = gas deviation factor at average flowing tubing pressure and temperature, fraction

Q = flow rate at standard conditions (at 14.65 psia and 60°F), MMSCF/D

d = I.D. of tubing, inches

The gas gravity ρ_g of equations of k and log F can be calculated from the composition of flowstream or measured experimentally. If composition is not available and experimental determination is not possible the gas gravity of composite flowstream may be obtained from the following equation:

$$\rho_g = G_G + \frac{4591 \, \rho/R}{1 + 1123/R}$$

where

ρ_g = full-stream gas (well-fluid) gravity

G_G = separator-gas gravity (dry)

ρ = specific gravity or well-fluid

R = gas-liquid ratio, cu ft/BBL

Using the data in table 5.3, a log-log plot of $(\bar{p}_R^2 - p_{wf}^2)$ was prepared and is shown in figure 5.2. From the plot, the AOF is read opposite $(\bar{p}_r^2 - p_{wf}^2) = p_e^2$ as 23 MMSCF/D.

The slope of the line or the tangent of its angle to the horizontal gives the inverse of n, that is,

$$\frac{1}{n} = \frac{\log(11.5 \times 10^6) - \log(1.1 \times 10^6)}{\log(10 \times 10^6) - \log(1 \times 10^6)} = 1.02$$

or $n = 0.980$.

And for the constant C, we can write

$$\text{AOF} = C(\bar{p}_R^2 - p_{wf}^2)^n$$
$$24 = C(27.668 \times 10^6)^{0.980}$$
$$C = 0.000001222 \text{ (MMSCF/D)/psi}^2$$

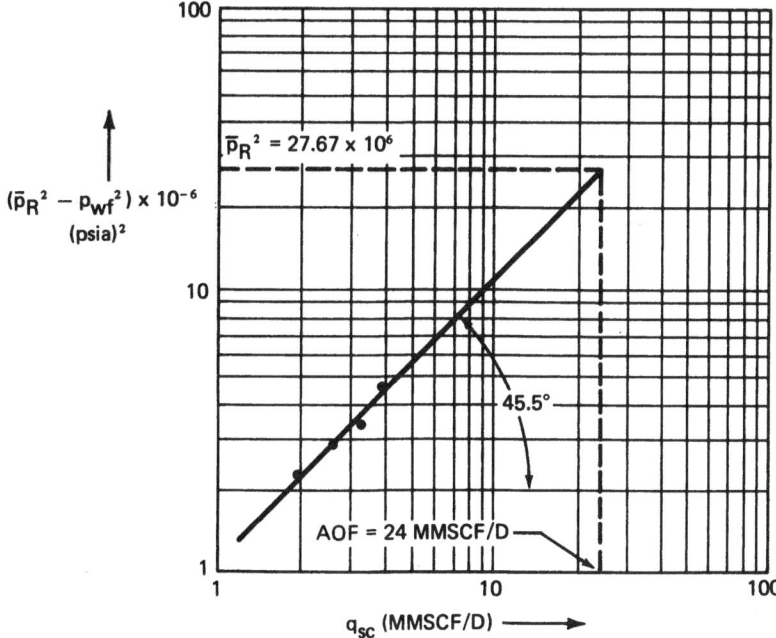

Figure 5.2
Plot of Deliverability Data for Melancon Heirs #1.

If the AOF used had been 24,000 MSCF/D instead of 24 MMSCF/D, then the value of C in the more appealing units would be $C = 0.001222$ (MSCF/D)/psi².

Chapter 6

Exercises

Exercise 1

1.1. Using the test data below and log-log paper find the values of n and C. What is the AOF of the well?

Stabilized Bottom Hole Pressure, psia	Producing Rate MMSCF/D
2800	0
2680	1.8
2590	2.7
2500	3.6
2425	4.5

Answer: AOF = 14.5 MMSCF/D or by calculating at $P_{wf}^2 = 0$, $q_{sc} = 14.64$ MMSCF/D.

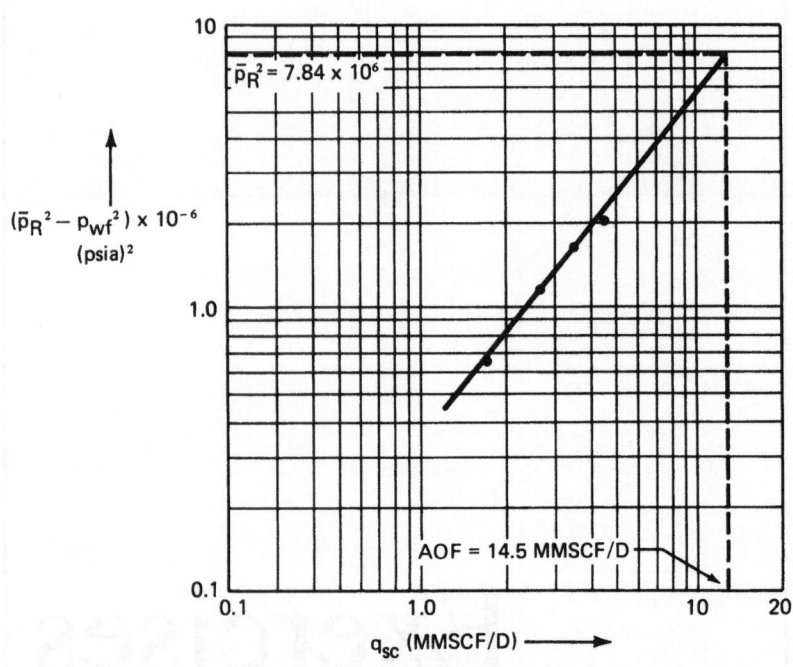

Figure 6.1
Graph of the Solution to Problem 1.1.

1.2. If the well tested in problem 1.1 delivers gas to a pipeline that has an operating pressure of 350 psia and the equivalent sandface pressure is 390 psia, what is the deliverability of this well assuming original reservoir pressure exists?

Answer: $q_{sc} = 14.4 \ MMSCF/D$.

1.3. Solve problem 1.2 as if the reservoir has been depleted to the point where the reservoir pressure has dropped to 1200 psia. Assume that C and n do not change significantly.

Answer: $q_{sc} = 3.4 \ MMSCF/D$.

Supplemental Exercises

1.4. Derive the conventional backpressure test equation.

1.5. What are the basic assumptions made in the derivation of this equation?

1.6. Draw the flow rate versus time and sandface pressure versus time for a conventional backpressure test.

1.7. Briefly state what you would do in a sequential manner at the well site if you were to conduct a conventional backpressure test.

1.8. When can you say that a well has stabilized?

1.9. What is the range of the values of n and what do the different values mean?

1.10. What does absolute open-flow potential mean?

Exercise 2

2.1. Analyze the following modified isochronal welltest data to determine *n*, *C*, and AOF.

Time of Run (Hours)	Sandface Pressure (psia)	Flow Rate (MMSCF/D)	Remarks
16	2200	0	Initial shut-in
12	2018	3.2	Flow 1
12	2178	0	Shut-in
12	1881	5.2	Flow 2
12	2160	0	Shut-in
12	1712	7.0	Flow 3
12	2138	0	Shut-in
12	1584	8.0	Flow 4
40	1352	8.0	Extended flow
83	2200	0	Final shut-in

Answer: n = 0.38, C = 0.000004547,
AOF = 12.6 MMSCF/D.

Supplemental Exercises

2.2. Draw the flow rate versus time and sandface pressure versus time for an isochronal well test.

2.3. Briefly state the stepwise procedure that you would follow at the well site if you were to conduct an isochronal well test.

2.4. Why is isochronal well testing preferable to the conventional backpressure test?

2.5. Following isochronal welltest data is given:

Time of Run (Hours)	Sandface Pressure (psia)	Flow Rate (MMSCF/D)	Shut-in BHP (psia)
	3000	0	3000
2	2910	1.9	3000
2	2860	2.7	3000
2	2804	3.6	3000
2	2750	4.5	3000
16	2630	4.5	

The well continued to produce at 4.5 MMSCF/D and reached a stabilized sandface pressure of 2630 psia. Find the values of *n* and *C* and also calculate AOF for this well.

2.6. Draw the flow rate versus time and sandface pressure versus time for a modified isochronal well test.

2.7. Under which conditions is the modified isochronal test preferable to the isochronal test?

Exercise 3

3.1. What is the superiority of pseudopressure treatment over pressure-squared treatment?
Answer: In the pseudopressure treatment, it is not necessary to make ideal gas assumptions with respect to the behavior of viscosity and compressibility factor as a function of pressure. This increases the accuracy of the calculations.

3.2. What are the three different states of flow observed when a well is producing at a constant rate from a closed reservoir? List them in the chronological order they appear. What is the characteristic of the last state of flow?
Answer: Three states of flow are observed in the following chronological order: (a) Unsteady state or transient. (b) Transitional or late transient.

(c) Pseudosteady state. Pseudosteady state flow is characterized by the linear change of pressure with time.

3.3. What is the importance of skin factor? What does a negative skin mean?
Answer: Skin factor characterizes the deviation from ideal flow performance in the vicinity of the wellbore. A positive skin may indicate the presence of a formation damage and/or limited entry, whereas a negative skin indicates improvement (stimulation).

3.4. What are the components of the composite skin factor, S'?
Answer: Components of the composite skin factor are physical skin and non-Darcy flow skin. Physical skin is caused by formation damage or formation stimulation and limited entry. Non-Darcy flow skin effect represents the additional pressure drop due to turbulent flow conditions around the wellbore and through gravel-packed perforations.

Supplemental Exercise

3.5. Using pseudopressures derive the partial differential equation describing the flow of gas in a radial-cylindrical porous medium. Hint: start with equation (2.15).

Exercise 4

4.1. The following pressure buildup data was obtained on a gas well located in the center of a circular reservoir.

Shut-in Time, Δt (Hours)	Well Pressure, p_{wf} (psia)
0	1700
1	1930
2	2170
5	2221
14	2270
22	2291
35	2308
45	2316
60	2328

Other data for this well are:

Stabilized gas production rate before shut-in	= 5 MMSCF/D
Reservoir temperature	= 180°F
Formation thickness	= 62 ft
Average gas viscosity	= 0.02 cp
Gas compressibility factor	= 0.82 (average)
Porosity	= 18%
Cumulative gas production	= 900 MMSCF
Compressibility	= 5×10^{-3} psi^{-1}
Well radius	= 4 inches

Determine the permeability and composite skin factor. What is the static reservoir pressure?
Answer: $k = 2.95$ md (A very tight formation),
$S' = 0.83$ indicates a small skin,
$\bar{p}_R^2 = 6.31 \times 10^6$ or $\bar{p}_R = 2512$ psia.

4.2. What is the additional pressure drop due to skin in exercise 4.1?
Answer: $\Delta p_{skin} = 97$ psia.
Since S' is positive in exercise 4.1, Δp calculated

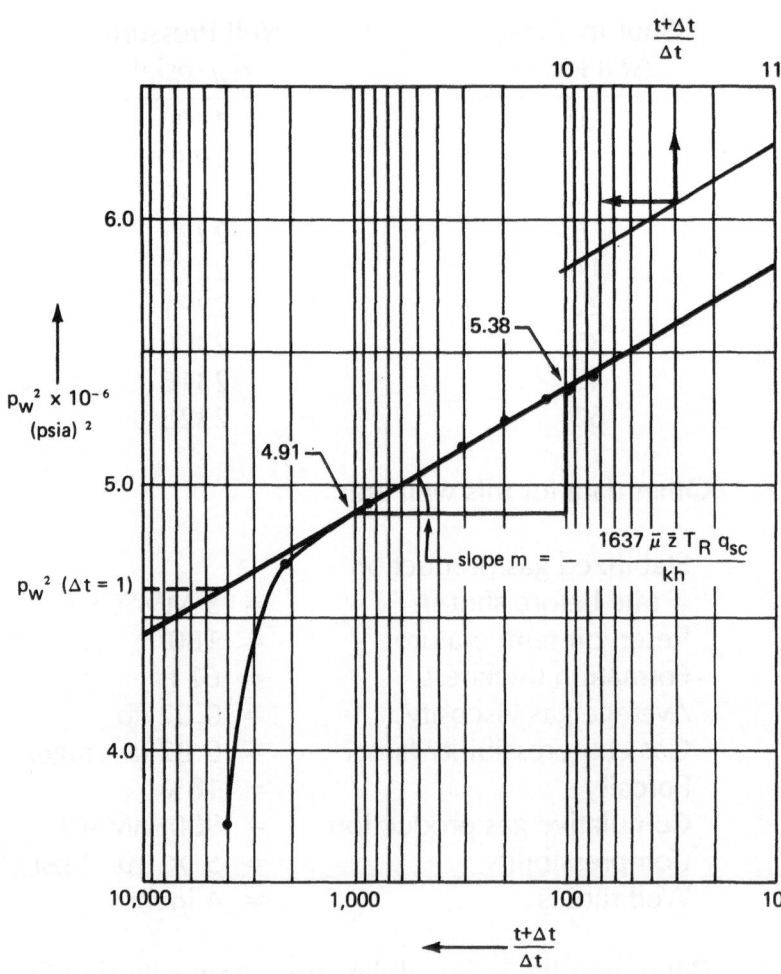

Figure 6.2
Graph of the Solution to Problem 4.1.

above is an additional pressure loss due to completion damage around the wellbore.

Supplemental Exercises

4.3. Composite skin values of 4.2 and 5.3 are obtained from two cycle pressure buildup-drawdown tests with flow rates of 3.4 MMSCF/D and 6.1 MMSCF/D, respectively. What is the magnitude of the physical skin for this specific well?

4.4. The following drawdown test data are obtained from a well in a cylindrical dry gas reservoir. The well was shut-in for a long period of time before the drawdown test was started. The well was then opened to production at a rate of 1 MMSCF/D.

Time (Hour)	p_{wf} (psia)
0.0	2000
0.1	1904
0.2	1898
0.3	1892
0.4	1888
0.6	1885
0.8	1882
0.9	1880
1.0	1879

Other relevant data are as follows:

Formation thickness   =   4 ft
Reservoir temperature   =   200°F
Porosity   =   20%
Gas gravity (air = 1.00) =   0.850

Find the permeability of the formation.

Exercise 5

5.1. (a). Is it important to use consistent pressure measurements (tophole or bottomhole) during the life of a gas well?

(b). Is it more critical in shallow or deep wells to use bottomhole pressures?

Answer: (a) It is very important to use consistent pressure measurements (tophole or bottomhole) during the life of a gas well. Correlations used in converting wellhead pressures to sandface pressures are not exact. Hence, consistency for comparative purposes is very important. (b) The use of bottomhole pressures is especially more desirable in deep wells. Since the gravity of gas is low, the pressure caused by a column of gas will be very small in shallow wells. In fact, the IOCC recommends the use of wellhead pressures as sandface pressures if the well depth is less than 2000 psia.

5.2. Calculate the flow rate from a well test including the condensate equivalent for a reservoir with the following production (see equations (3.12) and (3.13) and accompanying description):

Separator gas	= 15 MMSCF/D
Condensate production	= 150 STB/D
Specific gravity	= 0.759

Answer: total gas production = 15.12 MMSCF/D.

Supplemental Exercises

5.3. (a) If long stabilization times are expected, what type of drawdown test should be run?

(b) What is the usual cause of long stabilization times?

(c) Name four criteria used in selecting flow rates for a deliverability test.

5.4. What are wellbore storage effects?

5.5. Name two factors that determine the necessary duration of a flow test.

5.6. During a well test, the stabilized shut-in ($q_{sc} = 0$) tophole pressure was 2759 psia. If the following data apply, what is the bottomhole pressure? (See Cullender-Smith approach.)

Gas specific gravity   = 0.665
Well depth   = 6500 feet
Wellhead temperature   = 67°F
Formation temperature   = 213°F

5.7. If the well in exercise 5.6 was flowing 12.5 MMSCF/D (as measured by a critical flow prover) through 5.24-inch ID casing, how would you calculate the flowing bottomhole pressure?

Exercise 6

6.1 Given the following orifice meter configuration and data (use Appendix E):

Flange taps   11.938" ID pipe; 4" orifice
Upstream static pressure   = 692.5 psia
Orifice pressure drop   = 27.0 inches H_2O
Flowing temperature   = 75°F
Gas gravity   = 0.63
Base temperature   = 60°F
Base pressure   = 14.65 psia

(a) What are the minimum lengths of straight pipe needed upstream and downstream of the orifice?

(b) What is the flow rate in MMSCF/D?

Answer: (a) Upstream = 102 inches minimum before the orifice

Downstream = 37.2 inches minimum after the orifice

(b) q_{sc} = 14.179 MMSCF/D

Supplemental Exercises

6.2. Suppose that during a well test gas is flowing through a 2" critical flow prover with a 1/2" orifice. If the pressure in the line is 800 psia, temperature is 70°F, and the gas gravity is 0.65, what is the flow rate in MMSCF/D?

6.3. Under which conditions will the most complex well testing facilities be necessary?

6.4. What are the different ways of preventing hydrate formation?

6.5. What are the three basic types of bottomhole pressure gauges?

Exercise 7

7.1. What does the IOCC manual recommend if the value of n is greater than 1.0 or less than 0.50? What do you do if the same values of n are obtained in the second test?

Answer: If obtained values of n are greater than 1 or less than 0.500, IOCC recommends the retesting of the same well. If similar results are obtained from the retest data, then, a straight line is drawn from highest flow rate point with a slope of 1, if n is greater than 1. If n is less than 0.5, then, a straight line is drawn from the lowest flow rate point with a slope of 0.5.

Supplemental Exercises

7.2. How does the definition of stabilization time differ in Oklahoma and Texas? What is the definition in your area?

7.3. You are asked to design a backpressure test for a well that has already been producing. What specific procedures would you follow if the well is located in Oklahoma?

7.4. How would your procedure change if you were to test this well in Texas?, in Alberta? What is the procedure in your area?

Nomenclature & Units

The standard symbols adopted by the natural gas engineering industry are used throughout this manual. In the equations, practical rather than metric field units are used consistently. Definition of the symbols and associated units are presented in the following list.

A = area, sq ft
B_g = formation volume factor, cu ft/scf
C = gaswell performance constant, daily flow rate/psi^2
c = gas compressibility, psi^{-1}
c_{ws} = compressibility of wellbore fluids, psi^{-1}
D = turbulent flow factor, D/MSCF
d = depth, ft
d_i = internal diameter of casing or tubing, ft
F = friction factor, dimensionless
G = gas gravity (air = 1.00), dimensionless
$G.E.$ = gas equivalent, SCF/STB
h = formation thickness, ft
k = permeability, md
M = molecular weight, lb mol
m = mass, lb
p = pressure, psia
$\bar{p}_R$ = static reservoir pressure, psia
q_{sc} = gas flow rate, MMSCF/D or MSCF/D
R = ratio of dry gas to condensate, SCF/STB
r = radius, ft
S = skin factor, dimensionless
S' = composite skin factor, dimensionless
T_R = reservoir temperature, °R
t = time, hours or days
t_{ws} = wellbore storage time, hours
V = volume, cu ft
V_{ws} = volume of wellbore, cu ft
v = velocity, ft/D
z = compressibility factor, fraction

Greek Symbols

μ = viscosity, cp
ρ = density, lb/cu ft
ϕ = porosity, fraction
ψ = pseudopressure, psia/cp
γ = specific gravity

Subscripts

a = altered zone
cond   = condensation
e = external
g = gas
mf = middle flowing
r = radial direction
R = reservoir
sc = standard surface conditions (14.7 psia and 60°F)
tf = tophole flowing
w = well
wf = flowing well
ws = shut-in well or wellbore storage

Others

$-$ = average
$\triangle$ = difference
∂ = del operator

Appendix A

Derivation of Continuity Equation

In this section we will develop a mathematical statement of the continuity equation for the radial flow of gases through porous media (reservoir rock) into the wellbore. The equation of continuity (or the law of conservation of mass) states that for any given system

Rate of mass
accumulation $=$ (rate of mass in) $-$ (rate of mass out)

Consider a thin cylindrical shell of radius r, thickness (of reservoir rock) $\triangle r$ and length h as shown in figure A.1. In our analysis we shall assume one-dimensional flow through the outer face of the shell. The mass of gas in the shell is the gas density, ρ_g (mass/volume) times the porosity, ϕ, times the volume of the shell, $2\pi rh\triangle r$. The change in mass over the incremental time $\triangle t$,

$$(\phi\rho_g 2\pi rh\triangle r)_{t\,+\,\triangle t} - (\phi\rho_g 2\pi rh\triangle r)_t \qquad (A.1)$$

must equal the mass flowing during that time into the shell at r minus the mass flowing out at $r-\triangle r$. The mass flux (the

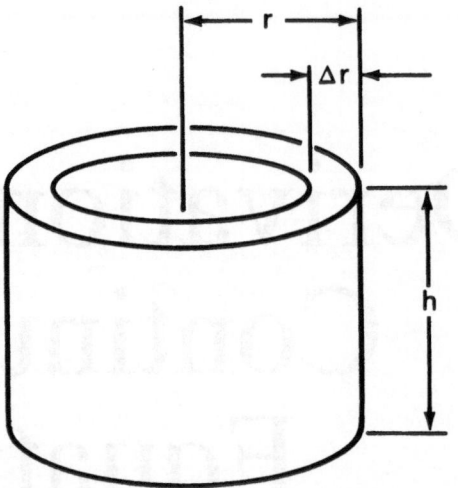

Figure A.1
Schematic for Derivation of Continuity Equation.

152   Gaswell Testing

mass flowing in the radial direction per unit surface area per unit time) is the radial velocity, v_r, times the mass density, ρ_g. The surface area at r is $2\pi rh$ while at $r - \Delta r$ it is $2\pi(r - \Delta r)h$. Conservation of mass for the gas dictates that

$$(\phi\rho_g 2\pi rh\Delta r)_{t + \Delta t} - (\phi\rho_g 2\pi rh\Delta r)_t = \Delta t \left\{ 2\pi rh\rho_g v_r \Big| - 2\pi(r - \Delta r)h\rho_g v_r \Big|_{r - \Delta r} \right\} \qquad (A.2)$$

Dividing by $2\pi rh\Delta r\Delta t$ gives

$$\frac{\phi\rho_g \big|_{t + \Delta} - \phi\rho_g \big|_t}{\Delta t} = \frac{r_g v_r \big|_r - r\rho_g v_r \big|_{r - \Delta r}}{r - \Delta r} \qquad (A.3)$$

Noting that $r \big|_{r - \Delta r} = r - \Delta r$ we may take the limits as Δr and $\Delta t \to 0$ and obtain

$$\frac{\partial}{\partial t}(\phi\rho_g) = -\frac{1}{r}\frac{\partial}{\partial r}(r\rho_g v_r) \qquad (A.4)$$

Appendix B

Analytical Solution of Diffusivity Equation

In this section we shall develop the analytical solution of the diffusivity equation, equation (2.17) which was developed using the pressure-squared approach. We start with repeating the derived diffusivity equation:

$$\frac{\partial^2 p^2}{\partial r^2} + \frac{1}{r}\frac{\partial p^2}{\partial r} = \frac{\phi\bar{\mu}_g}{kp}\frac{\partial p^2}{\partial t} \qquad (2.17)$$

The solution will be developed for the case of a well producing at a constant rate, q, from an infinite reservoir, the following boundary and initial conditions are associated with equation (2.17).

1.
Inner Boundary Condition

The flow rate, q, at the wellbore is constant, and q is positive for production. Then, from Darcy's law:

$$\left.\frac{qp}{\pi rh}\right|_{\text{well}} = \left.\frac{k}{\bar{\mu}_g}\frac{\partial p^2}{\partial r}\right|_{\text{well}}$$

$$\text{for } t>0; \quad \partial + r = r_w$$

or

$$\left.r\frac{\partial p^2}{\partial r}\right|_{\text{well}} = \frac{q\mu_g p}{\pi kh}\text{for } t>0; \partial + r = r_w \qquad (B.1)$$

If q is written in terms of standard conditions:

$$\left.r\frac{\partial p^2}{\partial r}\right|_{\text{well}} = \frac{q_{sc}\bar{\mu}_g}{\pi kh}\frac{p_{sc}T_R\bar{z}}{T_{sc}}$$

As a simplification which yields practically identical results, the well radius, r_w, is replaced by a line-sink. The boundary condition above then becomes:

$$\lim_{r \to 0} \quad r\frac{\partial p^2}{\partial r} \;=\; \frac{q_{sc}\bar{\mu}}{\pi kh}\;\frac{p_{sc}T_R\bar{z}}{T_{sc}}$$

$$\text{for } t>0; \;\; \partial + r \;=\; r_w \qquad (B.2)$$

2. Outer Boundary Condition

The pressure at the outer boundary (radius = infinity) is the same as the initial pressure for $t>0$. In other words, pressure approaches initial pressure as the radius approaches infinity.

$$p^2 \to p_r^2 \quad \text{as} \quad r \to \infty \quad \text{for } t \geq 0 \qquad (B.3)$$

3. Initial Condition

Initially, the pressure throughout the reservoir is constant, that is,

$$p^2 \;=\; p_r^2 \;(\partial + t \;=\; 0) \quad \text{for } r_w \leq r \leq r_e \qquad (B.4)$$

At this stage, we have completed the mathematical definition of the problem and, in doing so, we recognize that variables such as p, p_R, r, r_w, μ_g, k, h, ϕ, and t affect the solution of the flow equation.

Fundamental to the solution is the use of the Boltzmann transformation:

$$\chi = \frac{\phi \bar{\mu}_g}{pk} \frac{r^2}{4t} \quad \text{or} \quad \chi = \alpha \frac{r^2}{4t}$$

$$\text{if we let } \alpha = \frac{\phi \bar{\mu}}{pk} \quad \text{(B.5)}$$

to reduce the original partial differential equation to an ordinary differential equation.

Applying the chain rule, we can write the following:

$$\frac{\partial p^2}{\partial r} = \frac{\partial p^2}{\partial \chi} \frac{\partial \chi}{\partial r} \quad \text{(B.6)}$$

and

$$\frac{\partial p^2}{\partial t} = \frac{\partial p^2}{\partial \chi} \frac{\partial \chi}{\partial t} \quad \text{(B.7)}$$

and also,

$$\frac{\partial^2 p^2}{\partial r^2} = \frac{\partial}{\partial r} \left(\frac{\partial p^2}{\partial r} \right)$$

$$\frac{\partial^2 p^2}{\partial r^2} = \frac{\partial}{\partial \chi} \left(\frac{\partial p^2}{\partial r} \right) \frac{\partial \chi}{\partial r}$$

$$\frac{\partial^2 p^2}{\partial r^2} = \frac{\partial}{\partial \chi} \left(\frac{\partial p^2}{\partial \chi} \frac{\partial \chi}{\partial r} \right) \frac{\partial \chi}{\partial r} \quad \text{(B.8)}$$

From equation (B.5), the following equations can be obtained:

$$\frac{\partial \chi}{\partial r} = \alpha \frac{r}{2t} \quad \text{and} \quad \frac{\partial \chi}{\partial t} = -\alpha \frac{r^2}{4t^2} \quad \text{(B.9)}$$

Thus, equations (B.6), (B.7) and (B.8) are written using equation (B.9), such that,

$$\frac{\partial p^2}{\partial r} = \alpha \frac{r}{2t} \frac{\partial p^2}{\partial \chi} \quad \text{or} \quad \frac{\partial p^2}{\partial r} = \frac{2\chi}{r} \frac{\partial p^2}{\partial \chi} \quad \text{(B.10)}$$

$$\frac{\partial p^2}{\partial t} = \alpha \frac{r^2}{4t^2} \frac{\partial p^2}{\partial \chi} \quad \text{or}$$

$$\frac{\partial p^2}{\partial t} = -\frac{4\chi^2}{\alpha r^2} \frac{\partial p^2}{\partial \chi} \quad \text{(B.11)}$$

$$\frac{\partial^2 p^2}{\partial r^2} = \frac{\partial}{\partial \chi}\left(\frac{2\chi}{r} \frac{\partial p^2}{\partial \chi} \right) \frac{2\chi}{r}$$

$$\frac{\partial^2 p^2}{\partial r^2} = \frac{4\chi^2}{r^2} \frac{\partial^2 p^2}{\partial \chi^2} + \frac{\partial p^2}{\partial \chi}$$

$$\left[\frac{2r - 2\chi \left(\frac{\partial r}{\partial \chi}\right)}{r^2} \right] \frac{2\chi}{r}$$

or

$$\frac{\partial^2 p^2}{\partial r^2} = \frac{4\chi^2}{r^2} \frac{\partial^2 p}{\partial \chi^2} + \frac{\partial p^2}{\partial \chi} \left[\frac{2r - 2\chi \frac{r}{2\chi}}{r^2} \right] \frac{2\chi}{r}$$

making necessary cancellations:

$$\frac{\partial^2 p^2}{\partial r^2} = \frac{4\chi^2}{r^2} \frac{\partial^2 p^2}{\partial \chi^2} + \frac{2\chi}{r^2} \frac{\partial p^2}{\partial \chi} \quad \text{(B.12)}$$

Now equations (B.10), (B.11) and (B.12) can readily be substituted into the original equation, equation (2.17). Then,

$$\frac{4\chi^2}{r^2}\frac{\partial^2 p^2}{\partial \chi^2} + \frac{2\chi}{r^2}\frac{\partial p^2}{\partial \chi} + \frac{1}{r}\frac{2\chi}{r}\frac{\partial p^2}{\partial \chi} = -\frac{4\chi^2}{r^2}\frac{\partial p^2}{\partial \chi}$$

or gathering the similar terms:

$$\chi\frac{\partial^2 p^2}{\partial \chi^2} + (1+\chi)\frac{\partial p^2}{\partial \chi} = 0 \qquad \text{(B. 13)}$$

Equation (B.13) is the new form of the diffusivity equation after the application of the Boltzmann transformation. Note, that it contains only one independent variable and so may be written as an ordinary differential equation. Now, in order to be consistent we have to transform the boundary conditions too.

1.
Inner Boundary Condition:

$$\lim_{\chi \to 0} 2\chi \frac{\partial p^2}{\partial \chi} =$$

$$-\frac{q_{sc}\mu_g}{\pi kh}\frac{p_{sc}T_R\bar{z}}{T_{sc}} \qquad \text{for } 0 < \chi < \infty \qquad \text{(B.14)}$$

2.
Outer Boundary Condition:

$$p^2 \to p_R^2 \text{ as } \chi \to \infty \text{ for } t \geq 0 \qquad \text{(B.15)}$$

Let, $(p^2)^* = \dfrac{dp^2}{d\chi}$

Then, the ordinary differential equation of equation (B.13) becomes

$$\chi \frac{d(p^2)^*}{d\chi} + (1 + \chi)(p^2)^* = 0 \qquad \text{(B.16)}$$

Separation of the variables yields:

$$\frac{d(p^2)^*}{(p^2)^*} = -\frac{(1+\chi)}{\chi} d\chi \qquad \text{(B.17)}$$

Integrating equation (B.17), one obtains:

$$\ln (p^2)^* = -\ln \chi - \chi + C \qquad \text{(B.18)}$$

where C is an intermediate integration constant. Equation (B.18) can be written as:

$$(p^2)^* = \frac{C_1}{\chi} e^{-x} \qquad \text{(B.19)}$$

Using the first boundary condition, equation (B.14), the integration constant, C_1, can be evaluated:

$$-\frac{q_{sc}\bar{\mu}_g}{\pi kh} \frac{p_{sc}T_r\bar{z}}{pT_{sc}} \frac{1}{2\chi} = \frac{C_1}{\chi} e^{-x}$$

or as $\chi \to 0$

$$C_1 = -\frac{q_{sc}\bar{\mu}_g}{2\pi kh} \frac{p_{sc}T_R\bar{z}}{T_{sc}} \qquad \text{(B.20)}$$

Hence,

$$\frac{dp^2}{d\chi} = \frac{q_{sc}\bar{\mu}_g}{4\pi kh} \left[\frac{p_{sc}T_R\bar{z}}{T_{sc}} \right] \frac{1}{\chi} e^{-x} \quad \text{(B.21)}$$

Choosing the lower limit $\chi = \infty$, separating variables and integrating once more yields:

$$\int_{p_R}^{p} dp^2 = \frac{q_{sc}\bar{\mu}_g}{4\pi kh} \frac{p_{sc}T_R\bar{z}}{T_{sc}} \int_{\infty}^{\chi} \frac{e^{-x}}{\chi} d\chi \quad \text{(B.22)}$$

$$p^2 - p_R^2 = \frac{q_{sc}\mu_g}{4\pi kh} \frac{p_{sc}T_R\bar{z}}{T_{sc}} \int_{\infty}^{\chi} \frac{e^{-x}}{\chi} d\chi \quad \text{(B.23)}$$

But

$$\int_{\infty}^{\chi} \frac{e^{-x}}{\chi} d\chi = -\int_{\chi}^{\infty} \frac{e^{-x}}{\chi} d\chi$$

$$= - Ei(\chi)$$

where Ei (χ) is called the exponential integral and is shown graphically in figure B.1.

Equation (B.23) written for pressure p and radius r becomes:

$$p^2 = p_R^2 - \frac{q_{sc}\bar{\mu}_g}{4\pi kh} \frac{p_{sc}T_R\bar{z}}{T_{sc}} Ei (\chi) \quad \text{(B.24)}$$

It can be shown that for values of $\chi < 0.01$, Ei (χ) can be approximated as

$$Ei (\chi) \cong - \ln \chi - 0.5772$$

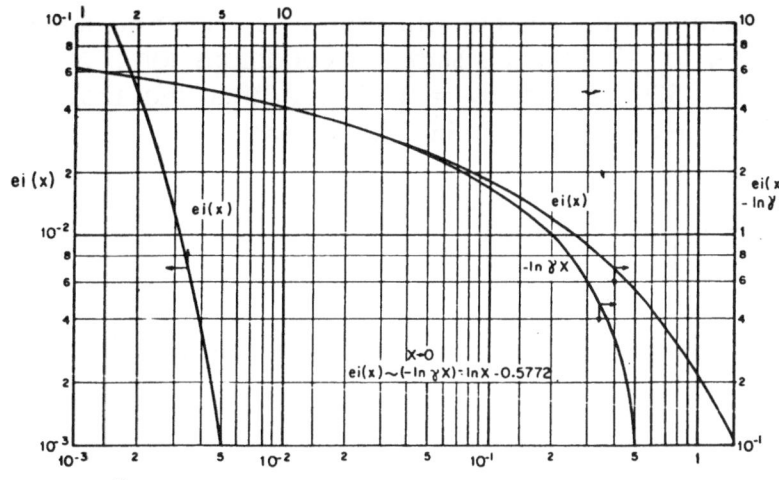

Figure B.1
Graph of the Exponential Integral. (From Dake, 1978.)

It is instructive to see when this approximation is valid. Normally the reservoir engineer is concerned with the measurement of pressure at the wellbore, that is where $r = r_w$. For this value of radius it can be seen that

$$\chi = \frac{\phi \bar{\mu}_g r^2_w}{4pkt}$$

will usually be less than 0.01 even for small values of t.

Equation (B.24) may be approximated with the following expression assuming that pressure is measured at the wellbore:

$$p^2_{wf} = p^2_R - \frac{q_{sc}\bar{\mu}_g}{4\pi kh}\left[\frac{p_{sc}T_R \bar{z}}{T_{sc}}\right]\left[\ln\frac{4pkt}{\phi\bar{\mu}_g r^2_w} - 0.5772\right] \quad \text{(B.25)}$$

Assuming that the compressibility, c, is approximately equal to $\frac{1}{p}$*, that the standard conditions for a gas are $T_{sc} = 520\ °R$ and $p_{sc} = 14.7$ psia, then the above expression can be written in field units as:

$$p^2_{wf} = p^2_R - \frac{1637\ q_{sc}\bar{\mu}_g z T_R}{kh}\left[\log t + \log\left(\frac{k}{\phi\bar{\mu}_g c r^2_w}\right) - 3.23\right]\qquad(B.26)$$

Note that in field units t is in hours and k is in millidarcies.

This substitution can be verified as follows:
The equation of state of a real gas where M is the molecular weight is given by:

$$\rho = \frac{M}{RT}\ \frac{p}{z}$$

For isothermal conditions

$$\frac{\partial\rho}{\partial p} = \frac{M}{RTz} + \frac{M}{RT}\ p\ \frac{\partial}{\partial p}\ (1/z)$$
or
$$\frac{\partial\rho}{\partial p} = \frac{\rho}{p} + \rho z\ \frac{\partial}{\partial p}\ (1/z)$$

Compressibility, c, of a gas is defined by $c = \frac{1}{\rho}\ \frac{\partial\rho}{\partial p}$. Substituting this definition into the previous equation one obtains

$$c = \frac{1}{p} - \frac{1}{z}\ \frac{\partial z}{\partial p}$$

if we assume ideal gas behavior, $z = 1$ and $\frac{\partial z}{\partial p} = 0$, then $c = \frac{1}{p}$

Appendix C

Skin Due to Restricted Entry

The effect of limited entry under conditions of Darcy flow is analyzed in detail by Odeh.* Wells that have been opened to flow along a fraction of their productive interval are termed wells with limited entry. It is obvious that limited entry to flow decreases well productivity. The magnitude of this decrease in well productivity will depend on the fraction of the formation open to flow, on the location of the open interval, on the ratio of drainage radius of the well to well radius, and on the thickness of the sand.

Procedure:

1. Calculate $Y_1 = \dfrac{\bar{z}_1}{H}$ where $\dfrac{H}{z_1} = \dfrac{\text{sand thickness, ft}}{\text{distance between the top of the productive interval and top of the open interval, ft}}$

2. Calculate the ratio of perforated interval to total thickness and find the proper chart (see figures C.1 to C.8).

3. Read the value of q/q_r from the chart using Y, and H. These charts are based on a drainage radius, r_e, of 660 ft and a well radius, r_w, of ¼ ft or $r_e/r_w = 2640$ ft. Use of these charts is left to your own discretion when there is a large variation in this ratio. However, a 20% change in the ratio results in about a 3% change in S_a.

---

*Odeh, A.S., "Steady-State Flow Capacity of Wells with Limited Entry to Flow," *Society of Petroleum Engineers Journal* (March 1968) 46-49; Trans., AIME, 243.

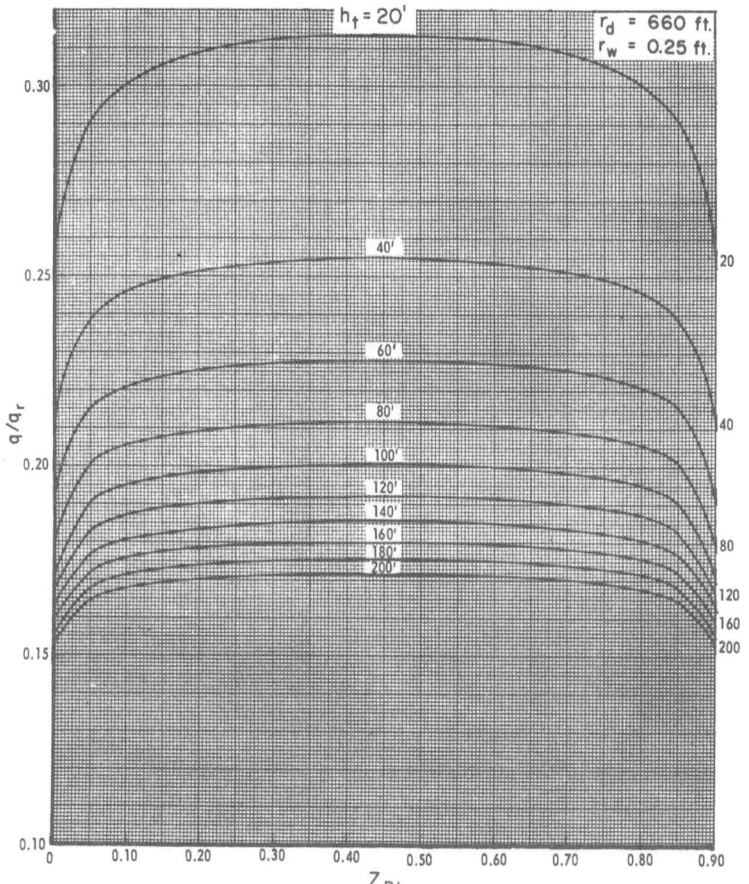

Figure C.1
Perforated Interval 0.1 of Thickness. (Copyright 1968, SPE-AIME. Odeh, A.S., "Steady-State Flow Capacity of Wells with Limited Entry to Flow," Society of Petroleum Engineers Journal (March 1968) 46-49; Trans., AIME, 243.)

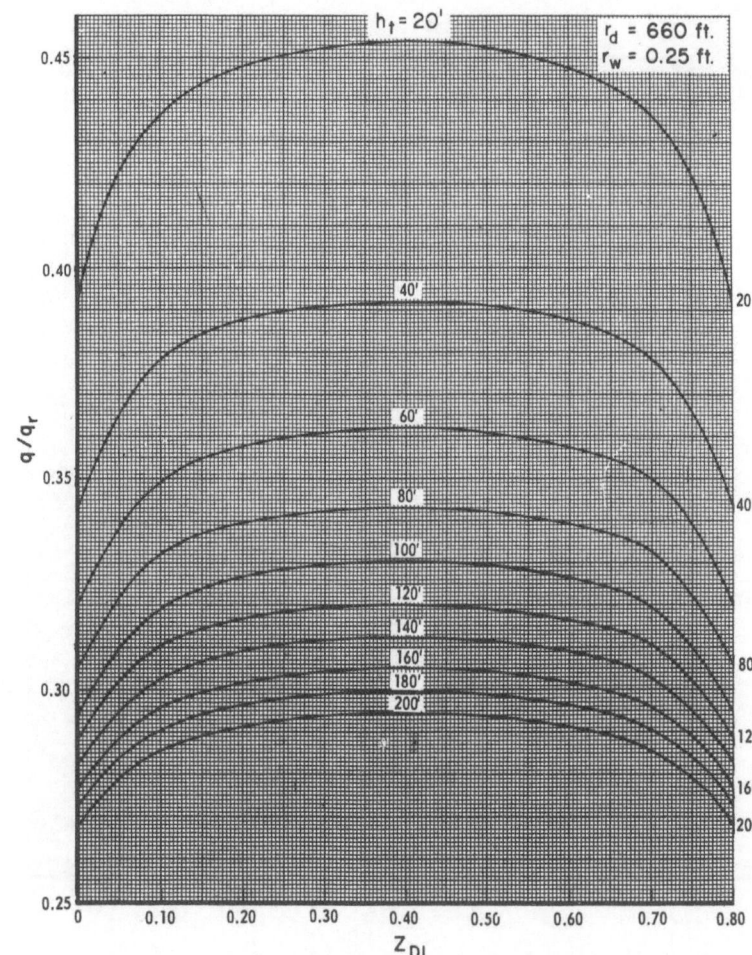

Figure C.2
Perforated Interval 0.2 of Thickness. (Copyright 1968, SPE-AIME. Odeh, A.S., "Steady-State Flow Capacity of Wells with Limited Entry to Flow," Society of Petroleum Engineers Journal (March 1968) 46-49; Trans., AIME, 243.)

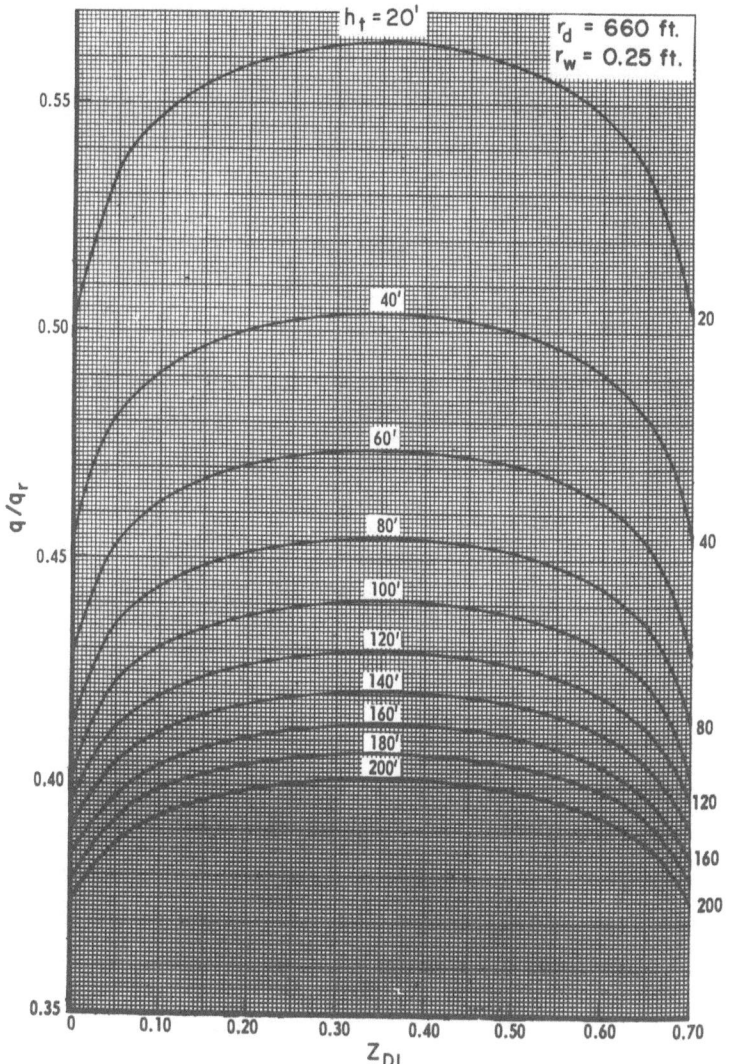

Figure C.3
Perforated Interval 0.3 of Thickness. (Copyright 1968, SPE-AIME. Odeh, A.S., "Steady-State Flow Capacity of Wells with Limited Entry to Flow," Society of Petroleum Engineers Journal *(March 1968) 46-49; Trans., AIME, 243.)*

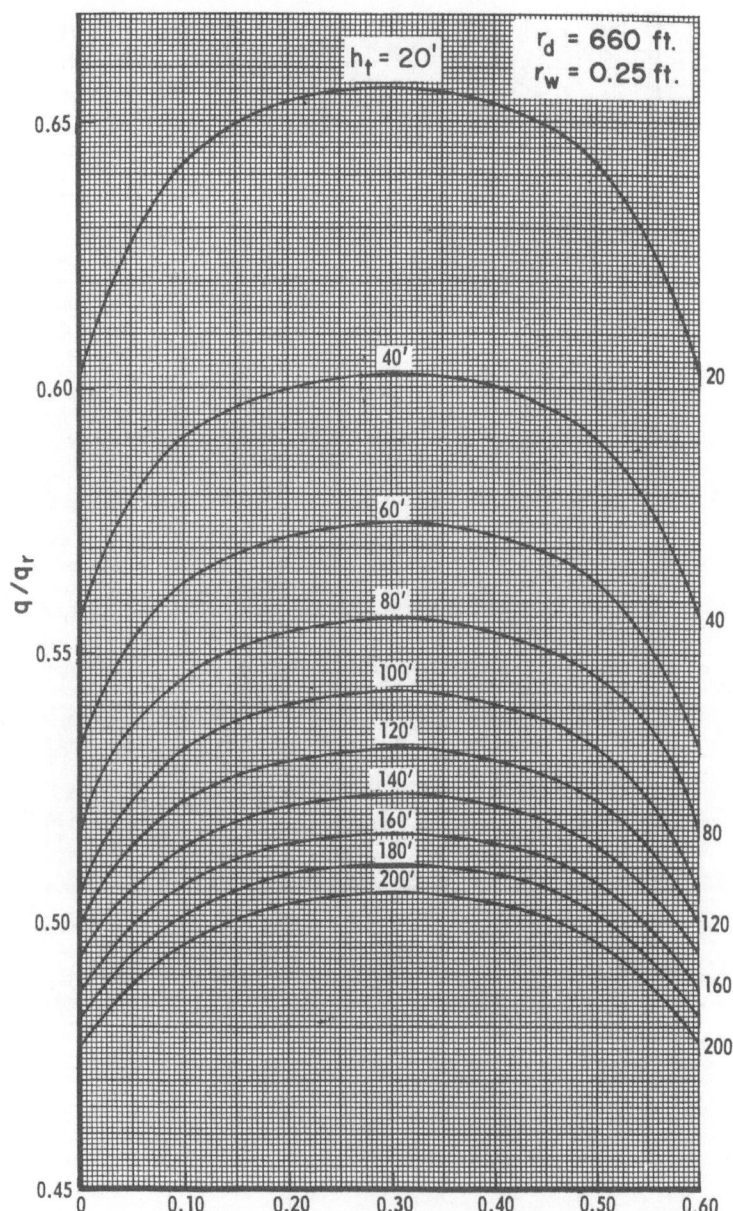

Figure C.4
Perforated Interval 0.4 of Thickness. (Copyright 1968, SPE-AIME. Odeh, A.S., "Steady-State Flow Capacity of Wells with Limited Entry to Flow," Society of Petroleum Engineers Journal *(March 1968) 46-49; Trans., AIME, 243.)*

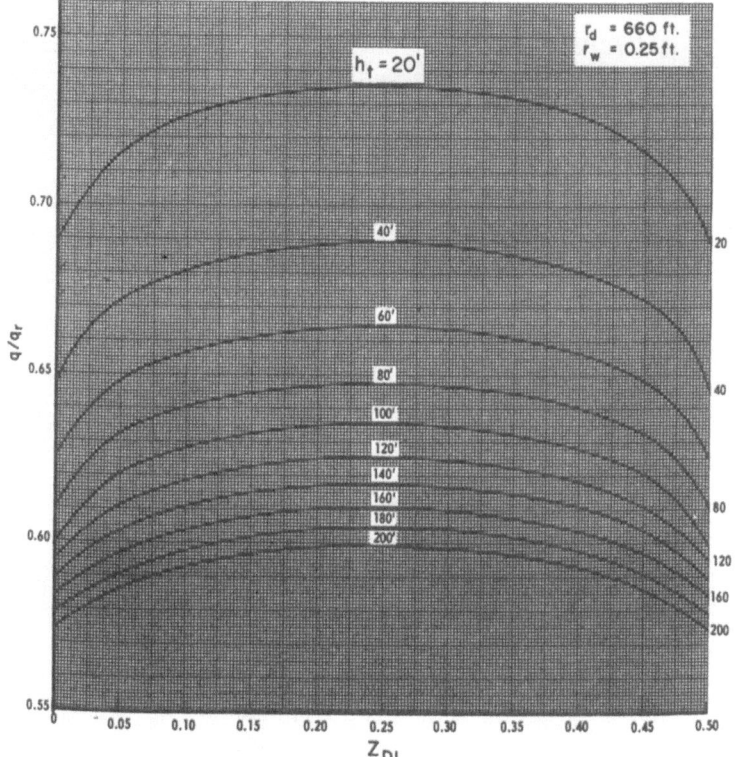

Figure C.5
Perforated Interval 0.5 of Thickness. (Copyright 1968, SPE-AIME. Odeh, A.S., "Steady-State Flow Capacity of Wells with Limited Entry to Flow," Society of Petroleum Engineers Journal (March 1968) 46-49; Trans., AIME, 243.)

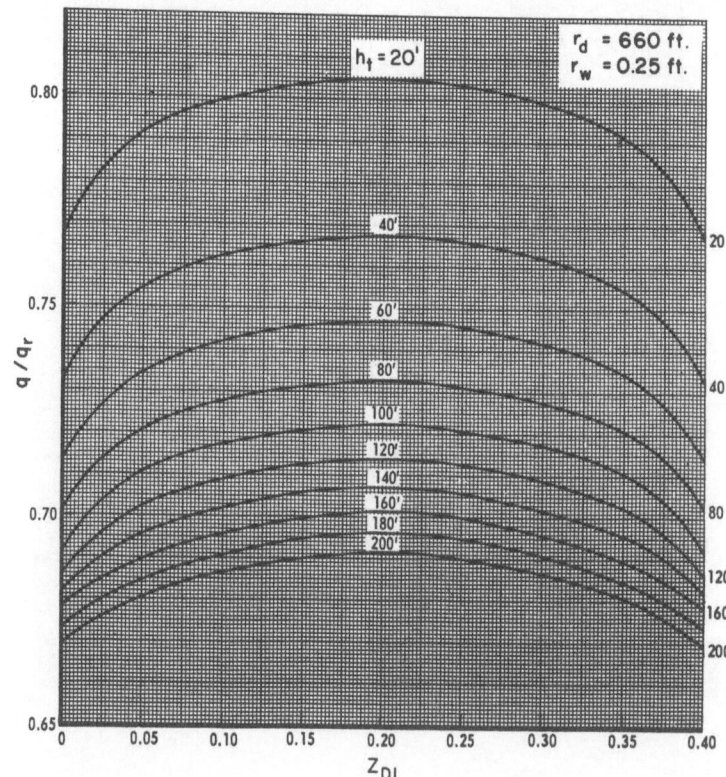

Figure C.6
Perforated Interval 0.6 of Thickness. (Copyright 1968, SPE-AIME. Odeh, A.S., "Steady-State Flow Capacity of Wells with Limited Entry to Flow," Society of Petroleum Engineers Journal (March 1968) 46-49; Trans., AIME, 243.)

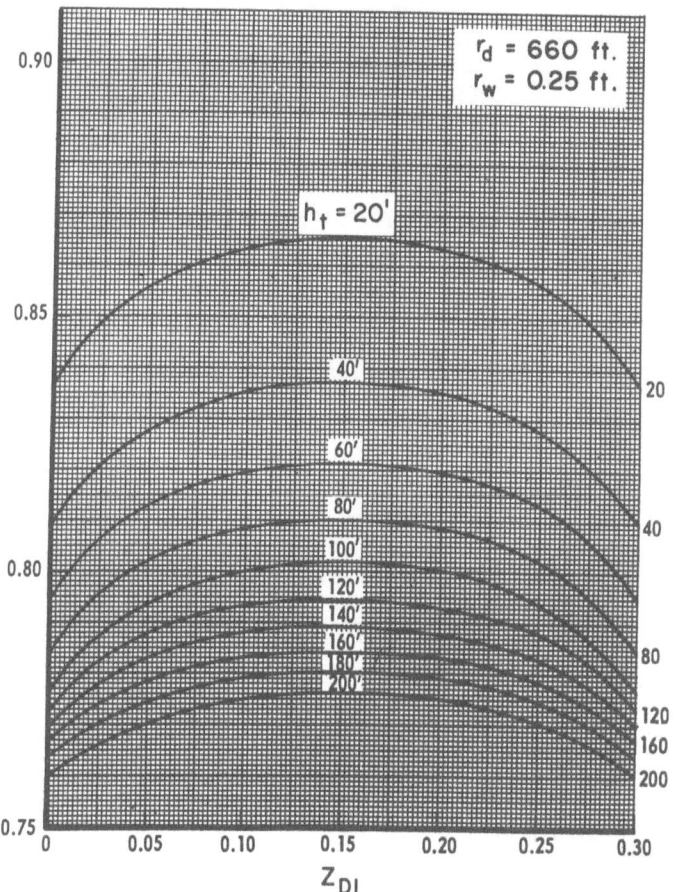

Figure C.7
Perforated Interval 0.7 of Thickness. (Copyright 1968, SPE-AIME. Odeh, A.S., "Steady-State Flow Capacity of Wells with Limited Entry to Flow," Society of Petroleum Engineers Journal (March 1968) 46-49; Trans., AIME, 243.)

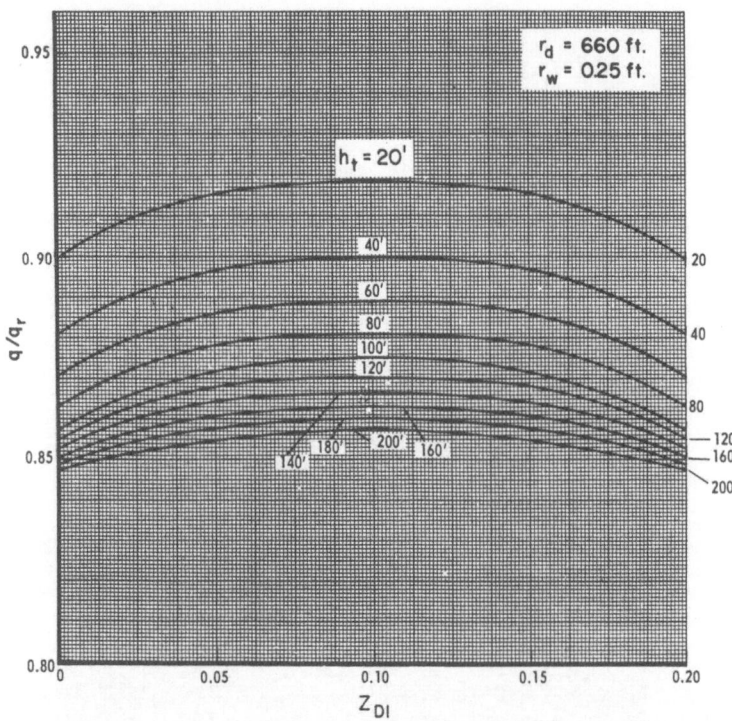

Figure C.8
Perforated Interval 0.8 of Thickness. (Copyright 1968, SPE-AIME. Odeh, A.S., "Steady-State Flow Capacity of Wells with Limited Entry to Flow," Society of Petroleum Engineers Journal (March 1968) 46-49; Trans., AIME, 243.)

4. Calculate apparent skin due to restricted entry from:

$$S_a = \frac{\left[\ln \dfrac{r_e}{r_w} - 0.75 \right] \left[1 - \dfrac{q}{q_r} \right]}{\dfrac{q}{q_r}}$$

where: r_e is the radius of drainage, ft
r_w is the well radius, ft

If one calculates the composite skin factor, S', from equation (2.42), it is possible to calculate the true skin, S_t, of the formation by using the relationship:

$$S_t = S^1 - S_a$$

S_a can also be found graphically using figure C.9. It should be noted that true skin, S_t, refers to that portion of the composite skin, S', which is due to permeability damage caused by invading solids or fluids.

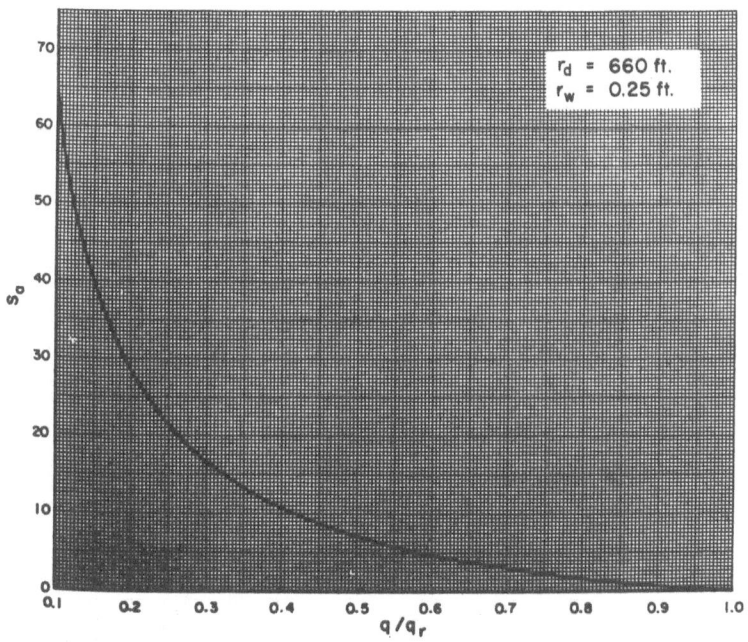

Figure C.9
q/q_r vs. s_a. (Copyright 1968, SPE-AIME. Odeh, A.S., "Steady-State Flow Capacity of Wells with Limited Entry to Flow," Society of Petroleum Engineers Journal (March 1968) 46-49; Trans., AIME, 243.)

Appendix D

Determination of Gas Supercompressibility Factor & Viscosity

Calculation of z, supercompressibility factor

To find a value of z at a given temperature and pressure we must:

(1) Calculate its critical temperature and pressure

(a) If gas gravity is given use figure D.1 to estimate p_{pc} and T_{pc} (where p_{pc} is the pseudocritical pressure, psia and T_{pc} is the pseudocritical temperature.

(b) If gas composition is known the critical constants for its individual components, P_{ci} and T_{ci} can be obtained from table D.1 and the following calculations are possible.

$$p_c = \sum_{i=1}^{n} mf_i \cdot p_{ci}$$

$$T_c = \sum_{i=1}^{n} mf_i \cdot T_{ci}$$

Where:

n = number of components in gas mixture
mf_i = mole fraction of the ith component
p_{ci} = critical pressure of the ith component, psia
T_{ci} = critical temperature of the ith component, °R
p_c = critical pressure of the mixture, psia
T_c = critical temperature of the mixture, °R

(2) Calculate pseudoreduced temperature and pressure using the following relationships

$$p_{pr} = p/p_{pc}$$
$$T_{pr} = T/T_{pc}$$

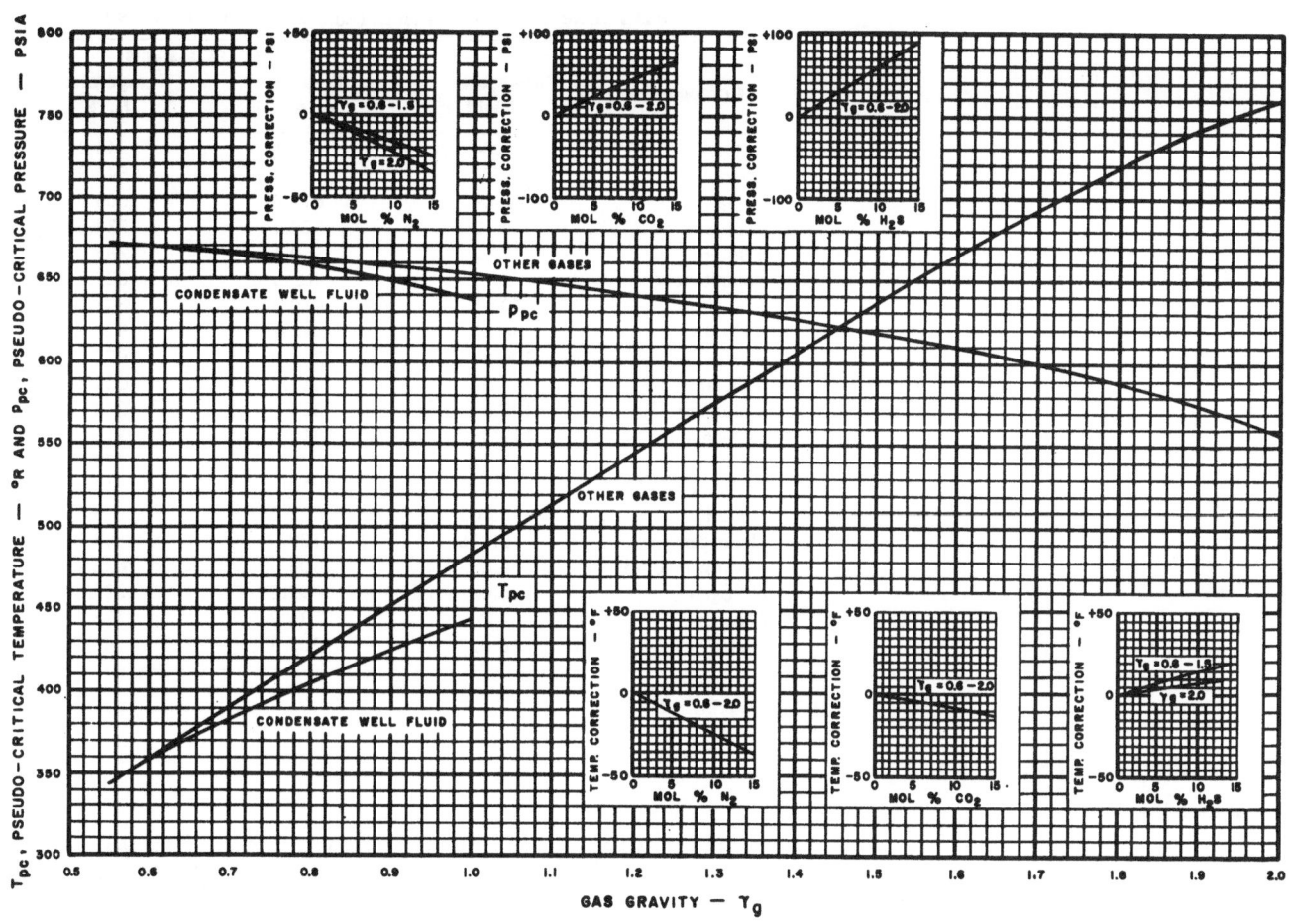

Figure D.1
Prediction of PseudoCritical Pressure and Temperature from Gas Gravity. (After Carr, Kobayashi and Burrows, Trans. AIME 201, *267, 1954. Modified by Mobil Oil Corporation.)*

where

p = pressure of the gas, psia
T = temperature of the gas, °R

(3) Knowing values for p_{pr} and T_{pr} use figure D.2 to find the value for z.

Table D.1 Table of Critical Constants. (Adapted from Burcik, 1979.)

Compound	Molecular Weight	Critical Constants	
		Pressure p_{cr} psia	Temperature T_{cr} °R
Methane	16.04	673.1	343.2
Ethane	30.07	708.3	549.9
Propane	44.09	617.4	666.0
Isobutane	58.12	529.1	734.6
n-Butane	58.12	550.1	765.7
Isopentane	72.15	483.5	829.6
n-Pentane	72.15	489.8	846.2
n-Hexane	86.17	440.1	914.2
n-Heptane	100.2	395.9	972.4
n-Octane	114.2	362.2	1024.9
n-Nonane	128.3	334	1073
n-Decane	142.3	312	1115
Air	28.97	547	239
Carbon dioxide	44.01	1070.2	547.5
Helium	4.003	33.2	9.5
Hydrogen	2.016	189.0	59.8
Hydrogen sulfide	34.08	1306.5	672.4
Nitrogen	28.02	492.2	227.0
Oxygen	32.00	736.9	278.6
Water	18.02	3209.5	1165.2

Calculation of μ, Gas Viscosity

(1) Knowing the molecular weight of the gas or its gravity (relative to air = 1.0) and the reservoir temperature in °F, μ_a, the viscosity of the gas at one atmosphere pressure may be found using figure D.3.

(2) Following the stepwise procedure used in estimating z factors, find the pseudoreduced temperature and pressure of the gas.

(3) With the estimated values of pseudoreduced temperature and pressure find the ratio, μ/μ_a, using figure D.4.

(4) Multiply μ/μ_a found in step 3 by μ_a, found in step 1, to find the viscosity of the gas at the given reservoir temperature and pressure.

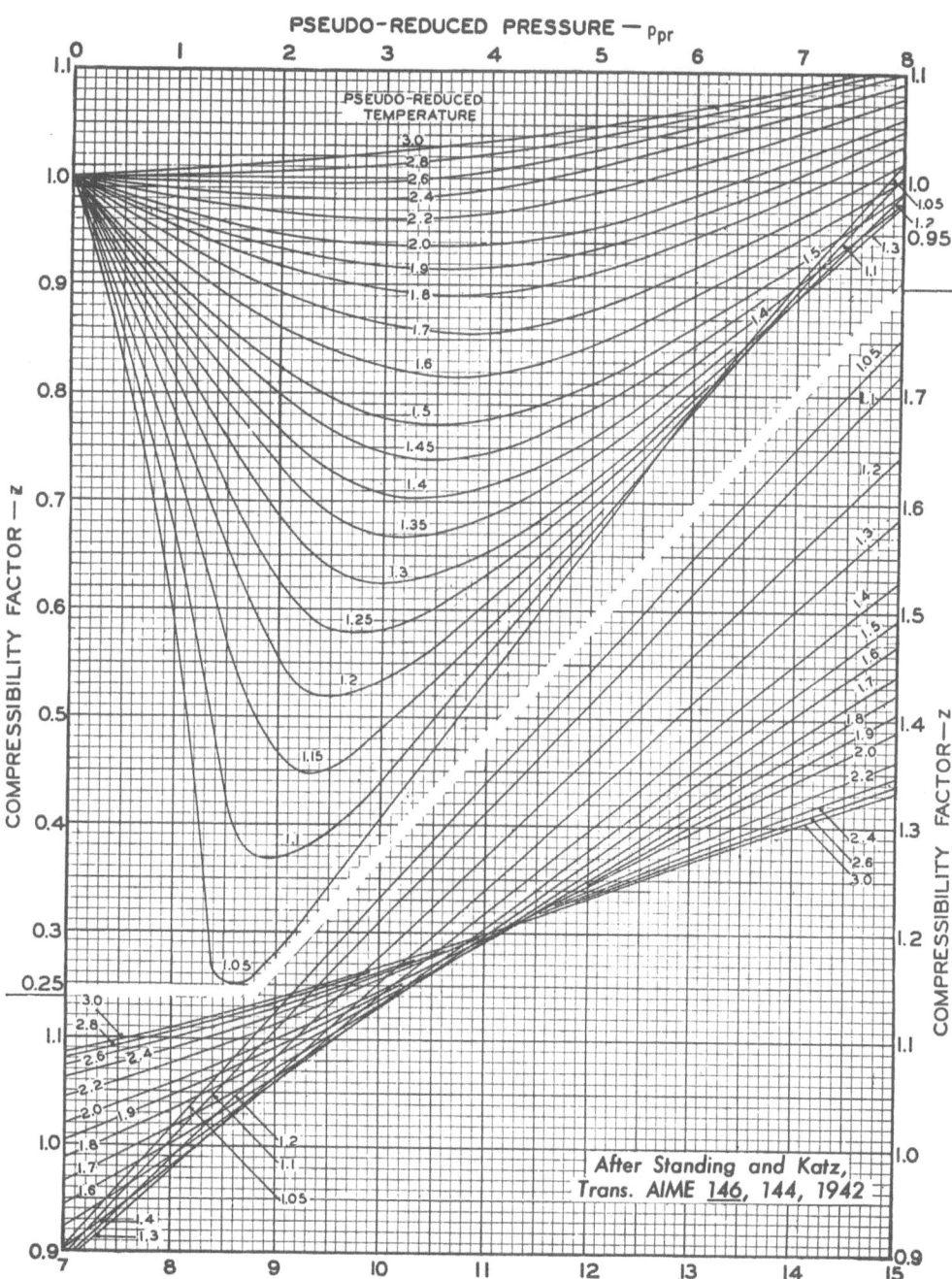

Figure D.2
Compressibility Factor for Natural Gases Versus PseudoReduced Pressure and Temperature. (After Standing and Katz, Trans. AIME 146, 144, 1942. Modified by Mobil Oil Corporation.)

Figure D.3

Viscosity of Paraffin Hydrocarbon Gases at One Atmosphere. (After Carr, Kobayashi and Burrows, Trans. AIME 201, 267, 1954. Modified by Mobil Oil Corporation.)

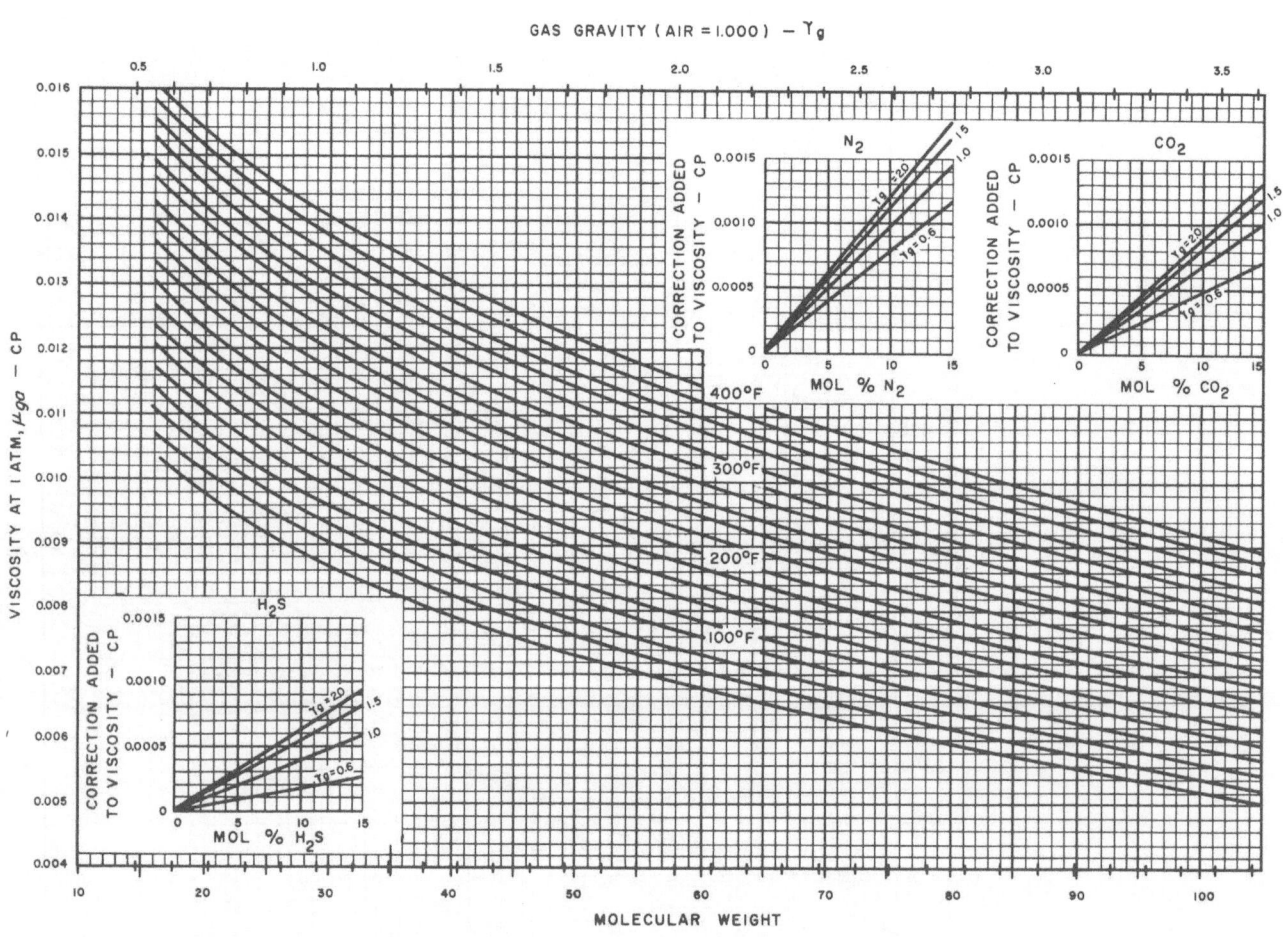

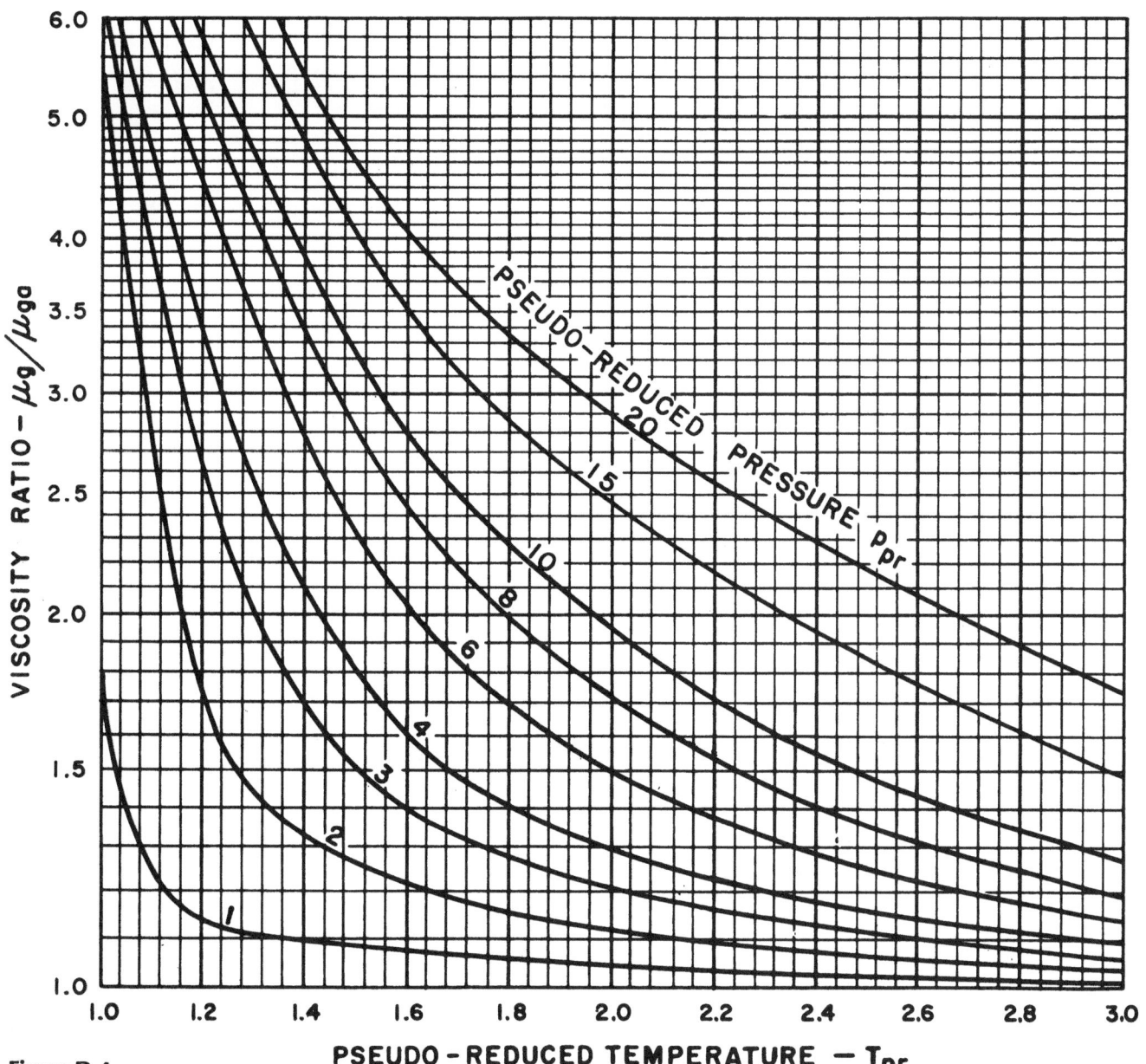

Figure D.4
*Viscosity Ratio Versus. Pseudo-Reduced
Temperature. (After Carr, Kobayashi and
Burrows, Trans. AIME 201, 267, 1954. Modified
by Mobil Oil Corporation.)*

Appendix | E

Orifice Coefficients

Tables and Figures reprinted by permission of the American Gas Association from *Orifice Metering of Natural Gas*, Arlington (1978).

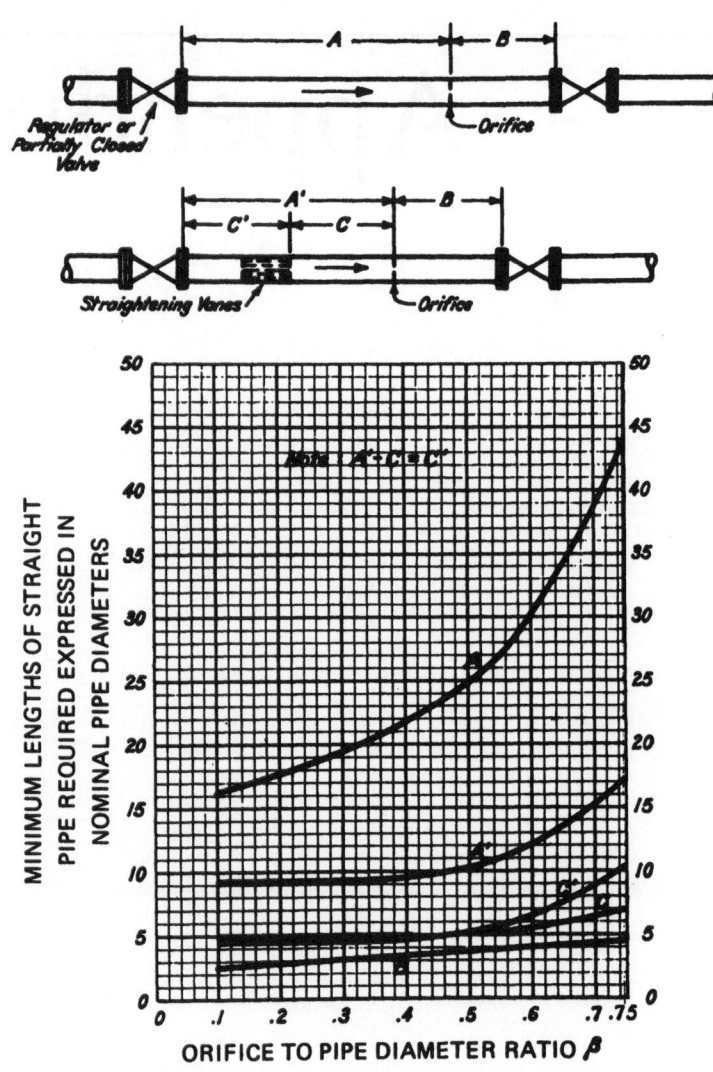

Figure E.1
Installation Sketch; Minimum Length of Straight Pipe Versus the Orifice to Pipe Diameter Ratio.

Note 1 When "Pipe Taps" are used, lengths A, A', and C shall be increased by 2 pipe diameters, and B by 8 pipe diameters.
Note 2 When the diameter or orifice may require changing to meet different conditions, the lengths of straight pipe should be those required for the maximum orifice to pipe diameter ratio that may be used.

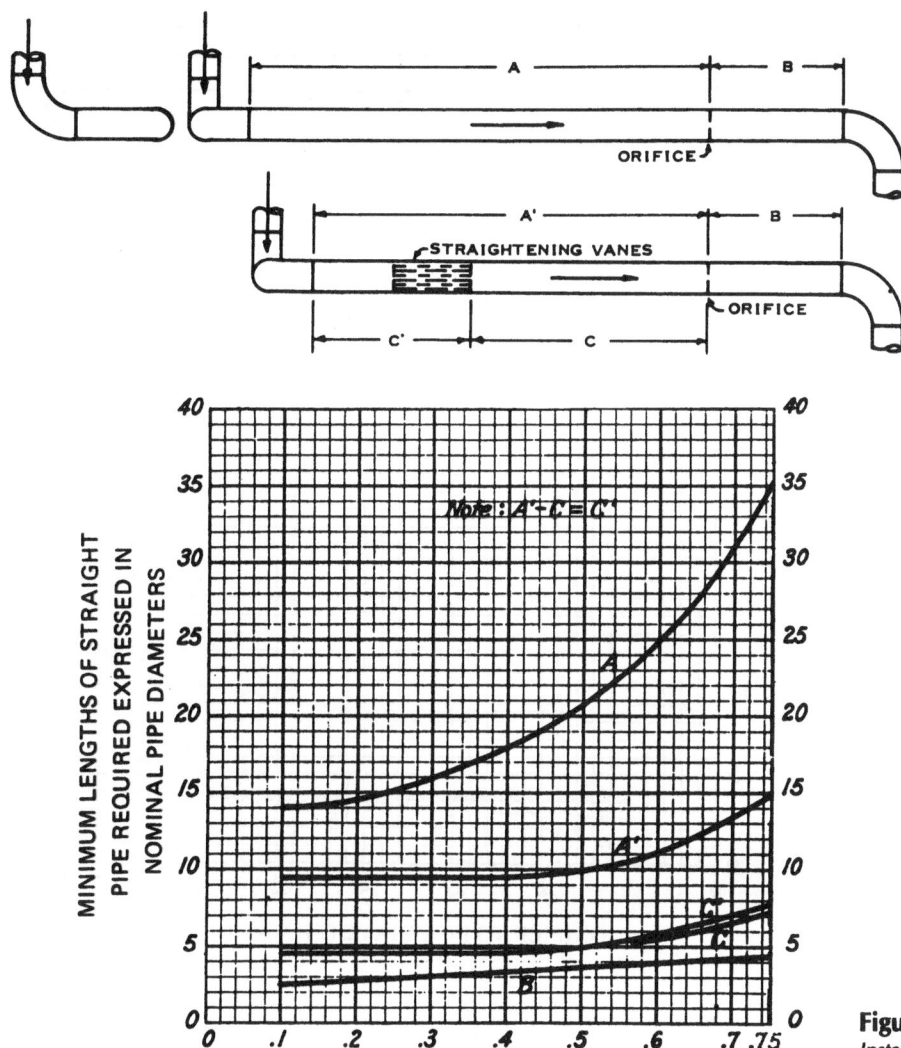

Figure E.2
Installation Sketch; Minimum Length of Straight Pipe Versus the Orifice to Pipe Diameter Ratio.

Note 1 When "Pipe Taps" are used, A, A', and C shall be increased by 2 pipe diameters, and B by 8 pipe diameters.

Note 2 When the diameter of the orifice may require changing to meet different conditions, the lengths of straight pipe should be those required for the maximum orifice to pipe diameter ratio that may be used.

Note 3 When the 2 ells shown in the above sketches are closely (less the [3D]) preceded by a third which is not in the same plane as the middle or second ell, the piping requirements shown by A should be doubled.

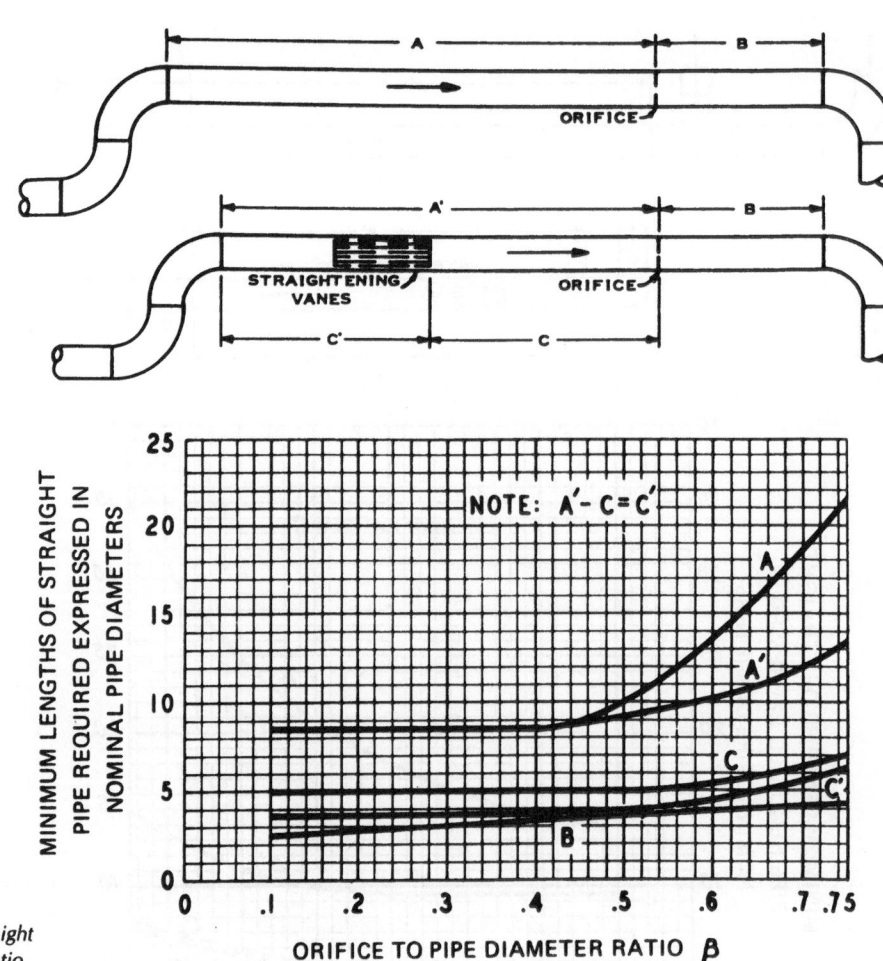

Figure E.3
*Installation Sketch; Minimum Length of Straight
Pipe Versus the Orifice to Pipe Diameter Ratio.*

Note 1 When "Pipe Taps" are used, A, A', and C shall be increased by 2 pipe diameters, and B by 8 pipe diameters.
Note 2 When the diameter of the orifice may require changing to meet different conditions, the lengths of straight pipe should be those required for the maximum orifice to pipe diameter ratio that may be used.

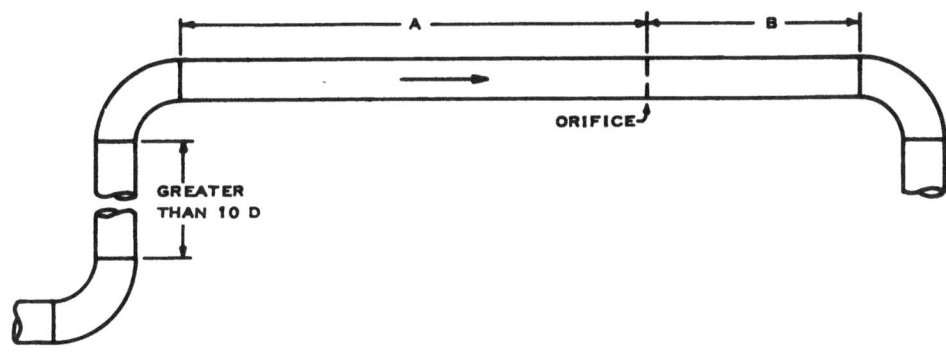

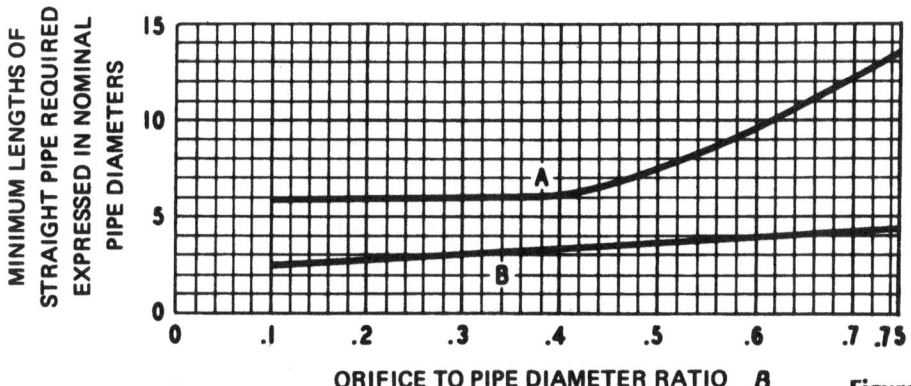

Figure E.4
Installation Sketch; Minimum Length of Straight Pipe Versus the Orifice to Pipe Diameter Ratio.

Note 1 When "Pipe Taps" are used, A shall be increased by 2 pipe diameters, and B by 8 pipe diameters.
Note 2 When the diameter of the orifice may require changing to meet different conditions, the lengths of straight pipe should be those required for the maximum orifice to pipe diameter ratio that may be used.
Note 3 The straight run of pipe between the elbows must be at least 10 diameters in length. If this length is less than 10 diameters figure E.3 shall be applicable.

Appendix E   **189**

TABLE E.1 F_b Basic Orifice Factors—Flange Taps

Base Temperature = 60° F   Flowing Temperature = 60° F $\sqrt{h_w\,p_f}$ = ∞
Base Pressure = 14.73 psia   Specific Gravity = 1.0 h_w/p_f = 0

Pipe Sizes — Nominal and Published Inside Diameters, Inches

Orifice Diameter Inches	2			3				4	
	1.689	1.939	2.067	2.300	2.626	2.900	3.068	3.152	3.438
.250	12.695	12.708	12.711	12.714	12.712	12.708	12.705	12.703	12.697
.375	28.474	28.440	28.427	28.411	28.393	28.382	28.376	28.373	28.364
.500	50.777	50.587	50.521	50.435	50.356	50.313	50.292	50.283	50.258
.625	80.090	79.508	79.311	79.052	78.817	78.687	78.625	78.599	78.523
.750	117.09	115.62	115.14	114.52	113.99	113.70	113.56	113.50	113.33
.875	162.95	159.56	158.47	157.12	156.00	155.41	155.14	155.03	154.71
1.000	219.77	212.47	210.22	207.44	205.18	204.04	203.54	203.33	202.75
1.125	290.99	276.19	271.70	266.35	262.06	259.95	259.04	258.65	257.63
1.250	385.78	353.58	345.13	335.12	327.39	323.63	322.03	321.37	319.61
1.375		448.59	433.50	415.75	402.18	395.80	393.09	391.97	389.03
1.500			542.27	510.86	487.98	477.36	472.96	471.14	466.39
1.625				623.91	586.82	569.65	562.58	559.72	552.31
1.750					701.26	674.44	663.42	658.96	647.54
1.875					834.88	793.88	777.18	770.44	753.17
2.000						930.65	906.01	896.06	870.59
2.125						1091.2	1052.5	1038.1	1001.4
2.250							1223.2	1199.9	1147.7
2.375									1311.7
2.500									1498.4

Orifice Diameter Inches	4		6				8		
	3.826	4.026	4.897	5.189	5.761	6.065	7.625	7.981	8.071
.250	12.688	12.683							
.375	28.353	28.348							
.500	50.234	50.224	50.197	50.191	50.182	50.178			
.625	78.450	78.421	78.338	78.321	78.296	78.287			
.750	113.15	113.08	112.87	112.82	112.75	112.72			
.875	154.40	154.27	153.88	153.78	153.63	153.56	153.34	153.31	153.31
1.000	202.20	201.99	201.34	201.19	200.96	200.85	200.46	200.40	200.38
1.125	256.69	256.33	255.31	255.08	254.72	254.56	253.99	253.89	253.87
1.250	318.03	317.45	315.83	315.48	314.95	314.72	313.91	313.78	313.74
1.375	386.45	385.51	382.99	382.47	381.70	381.37	380.25	380.06	380.00
1.500	462.27	460.79	456.93	456.16	455.03	454.57	453.02	452.78	452.72
1.625	545.89	543.61	537.77	536.64	535.03	534.38	532.27	531.95	531.87
1.750	637.83	634.39	625.73	624.09	621.79	620.89	618.02	617.60	617.50
1.875	738.75	733.68	721.03	718.69	715.44	714.18	710.32	709.77	709.64
2.000	849.41	842.12	823.99	820.68	816.13	814.41	809.22	808.50	808.34
2.125	970.95	960.48	934.97	930.35	924.07	921.71	914.78	913.85	913.64
2.250	1104.7	1089.9	1054.4	1048.1	1039.5	1036.3	1027.1	1025.9	1025.6
2.375	1252.1	1231.7	1182.9	1174.2	1162.6	1158.3	1146.2	1144.7	1144.3
2.500	1415.0	1387.2	1320.9	1309.3	1293.8	1288.2	1272.3	1270.3	1269.8
2.625	1595.6	1558.2	1469.2	1453.9	1433.5	1426.0	1405.4	1402.9	1402.3
2.750	1797.1	1746.7	1628.9	1608.7	1582.0	1572.3	1545.7	1542.5	1541.8
2.875		1955.5	1801.0	1774.5	1740.0	1727.5	1693.4	1689.3	1688.4
3.000		2195.0	1986.6	1952.4	1907.8	1891.9	1848.6	1843.5	1842.3
3.125			2187.2	2143.4	2086.4	2066.1	2011.6	2005.2	2003.8
3.250			2404.2	2348.8	2276.5	2250.8	2182.6	2174.6	2172.9
3.375			2639.5	2569.8	2479.1	2446.8	2361.8	2352.0	2349.9
3.500			2895.5	2808.1	2695.1	2654.9	2549.7	2537.7	2535.0
3.625			3180.8	3065.3	2925.7	2876.0	2746.5	2731.8	2728.6
3.750				3345.5	3172.1	3111.2	2952.6	2934.8	2930.8
3.875				3657.7	3435.8	3361.5	3168.3	3146.9	3142.1
4.000					3718.2	3628.2	3394.3	3368.5	3362.9
4.250					4354.8	4216.6	3879.4	3842.3	3834.2
4.500						4900.9	4412.8	4360.5	4349.0
4.750							5000.7	4928.1	4912.2
5.000							5650.0	5551.1	5529.5
5.250							6369.3	6236.4	6207.3
5.500							7170.9	6992.0	6953.6
5.750								7830.0	7777.8
6.000									8707.0

TABLE E.1 (Continued) F_b Basic Orifice Factors—Flange Taps

Base Temperature = 60° F Flowing Temperature = 60° F $\sqrt{h_w\ p_f}$ = ∞
Base Pressure = 14.73 psia Specific Gravity = 1.0 h_w/p_f = 0

Pipe Sizes —Nominal and Published Inside Diameters, Inches

Orifice Diameter Inches	10			12			16		
	9.564	10.020	10.136	11.376	11.938	12.090	14.688	15.000	15.250
1.000	200.20								
1.125	253.56	253.48	253.47						
1.250	313.31	313.20	313.18	312.94	312.85	312.83			
1.375	379.44	379.29	379.26	378.94	378.82	378.79			
1.500	451.95	451.76	451.72	451.30	451.14	451.10	450.53	450.48	
1.625	530.87	530.63	530.57	530.04	529.84	529.78	529.06	528.99	528.94
1.750	616.20	615.90	615.83	615.16	614.91	614.84	613.94	613.85	613.78
1.875	707.98	707.60	707.51	706.68	706.36	706.28	705.18	705.07	704.99
2.000	806.23	805.76	805.65	804.61	804.23	804.13	802.78	802.65	802.55
2.125	910.97	910.38	910.24	908.98	908.51	908.39	906.77	906.61	906.49
2.250	1022.2	1021.5	1021.3	1019.8	1019.2	1019.1	1017.1	1017.0	1016.8
2.375	1140.1	1139.2	1139.0	1137.1	1136.4	1136.2	1133.9	1133.7	1133.5
2.500	1264.5	1263.4	1263.1	1260.8	1260.0	1259.8	1257.1	1256.8	1256.6
2.625	1395.6	1394.2	1393.9	1391.1	1390.1	1389.9	1386.7	1386.4	1386.1
2.750	1533.4	1531.7	1531.3	1528.0	1526.8	1526.5	1522.7	1522.4	1522.1
2.875	1678.0	1675.9	1675.4	1671.4	1670.0	1669.6	1665.2	1664.8	1664.5
3.000	1829.4	1826.9	1826.3	1821.4	1819.7	1819.3	1814.1	1813.7	1813.3
3.125	1987.8	1984.7	1984.0	1978.1	1976.1	1975.6	1969.6	1969.0	1968.6
3.250	2153.2	2149.5	2148.6	2141.5	2139.2	2138.6	2131.5	2130.9	2130.4
3.375	2325.7	2321.2	2320.2	2311.7	2308.9	2308.2	2299.9	2299.2	2298.7
3.500	2505.6	2500.1	2498.9	2488.7	2485.4	2484.6	2474.9	2474.1	2473.5
3.625	2692.8	2686.2	2684.7	2672.6	2668.7	2667.7	2656.4	2655.5	2654.8
3.750	2887.6	2879.7	2877.9	2863.5	2858.8	2857.7	2844.6	2843.5	2842.7
3.875	3090.1	3080.7	3078.5	3061.4	3055.9	3054.6	3039.4	3038.1	3037.2
4.000	3300.6	3289.3	3286.8	3266.4	3260.0	3258.5	3240.8	3239.4	3238.3
4.250	3746.1	3730.2	3726.7	3698.4	3689.6	3687.5	3663.8	3661.9	3660.5
4.500	4226.0	4204.1	4199.2	4160.4	4148.4	4145.5	4113.9	4111.5	4109.7
4.750	4742.7	4712.8	4706.2	4653.4	4637.2	4633.4	4591.5	4588.4	4586.0
5.000	5298.6	5258.4	5249.6	5179.0	5157.4	5152.3	5097.2	5093.1	5090.1
5.250	5897.4	5843.6	5831.8	5738.5	5710.0	5703.3	5631.4	5626.1	5622.2
5.500	6543.1	6471.9	6456.3	6333.8	6296.6	6287.9	6194.8	6188.1	6183.1
5.750	7240.0	7146.9	7126.5	6966.9	6919.0	6907.8	6788.1	6779.6	6773.3
6.000	7993.3	7872.9	7846.6	7640.4	7579.0	7564.7	7412.3	7401.5	7393.6
6.250	8808.9	8654.8	8621.1	8357.3	8278.9	8260.7	8068.3	8054.8	8044.8
6.500	9693.3	9498.1	9455.3	9121.0	9021.7	8998.7	8757.3	8740.3	8727.9
6.750	10654	10409	10355	9935.2	9810.5	9781.6	9480.4	9459.4	9444.0
7.000	11711	11394	11327	10804	10649	10613	10239	10213	10194
7.250		12467	12381	11732	11540	11496	11035	11003	10980
7.500		13656	13541	12725	12489	12434	11869	11831	11803
7.750				13787	13500	13433	12745	12698	12664
8.000				14927	14578	14498	13664	13607	13566
8.250				16158	15730	15633	14628	14560	14510
8.500				17505	16963	16845	15642	15560	15500
8.750					18297	18148	16706	16609	16538
9.000						19566	17826	17711	17627
9.250							19004	18868	18769
9.500							20245	20085	19969
9.750							21552	21365	21229
10.000							22930	22712	22554
10.250							24385	24132	23947
10.500							25924	25628	25414
10.750							27567	27210	26960
11.000							29331	28899	28598
11.250								30710	30346

TABLE E.1 (Continued) F_b Basic Orifice Factors—Flange Taps

Base Temperature = 60° F Flowing Temperature = 60° F $\sqrt{h_w \, p_f}$ = —
Base Pressure = 14.73 psia Specific Gravity = 1.0 h_w/p_f = 0

Pipe Sizes — Nominal and Published Inside Diameters, Inches

Orifice Diameter Inches	20			24			30		
	18.814	19.000	19.250	22.626	23.000	23.250	28.628	29.000	29.250
2.000	801.40	801.35	801.29						
2.125	905.11	905.05	904.98						
2.250	1015.2	1015.1	1015.0						
2.375	1131.6	1131.5	1131.4	1130.2	1130.1	1130.0			
2.500	1254.4	1254.3	1254.2	1252.8	1252.6	1252.6			
2.625	1383.6	1383.5	1383.3	1381.7	1381.5	1381.4			
2.750	1519.1	1519.0	1518.8	1517.0	1516.8	1516.7			
2.875	1661.0	1660.9	1660.7	1658.6	1658.4	1658.3	1656.0		
3.000	1809.4	1809.2	1809.0	1806.6	1806.4	1806.2	1803.7	1803.5	1803.4
3.125	1964.1	1963.9	1963.7	1961.0	1960.7	1960.6	1957.7	1957.5	1957.4
3.250	2125.3	2125.1	2124.8	2121.7	2121.5	2121.3	2118.0	2117.9	2117.7
3.375	2292.9	2292.6	2292.3	2288.9	2288.6	2288.4	2284.7	2284.5	2284.4
3.500	2466.9	2466.6	2466.3	2462.4	2462.1	2461.8	2457.8	2457.6	2457.4
3.625	2647.3	2647.0	2646.6	2642.4	2642.0	2641.7	2637.3	2637.0	2636.8
3.750	2834.2	2833.9	2833.5	2828.7	2828.3	2828.0	2823.1	2822.8	2822.6
3.875	3027.5	3027.3	3026.8	3021.5	3021.0	3020.7	3015.2	3014.9	3014.7
4.000	3227.5	3227.1	3226.5	3220.6	3220.1	3219.8	3213.8	3213.5	3213.3
4.250	3646.7	3646.2	3645.6	3638.3	3637.7	3637.2	3630.1	3629.7	3629.4
4.500	4092.1	4091.5	4090.6	4081.8	4081.0	4080.5	4071.9	4071.4	4071.1
4.750	4563.7	4562.9	4561.9	4551.1	4550.1	4549.5	4539.3	4538.8	4538.4
5.000	5061.8	5060.8	5059.6	5046.4	5045.2	5044.5	5032.4	5031.8	5031.4
5.250	5586.6	5585.4	5583.8	5567.7	5566.4	5565.5	5551.3	5550.5	5550.0
5.500	6138.2	6136.7	6134.8	6115.3	6113.6	6112.6	6095.8	6094.9	6094.4
5.750	6717.1	6715.2	6712.8	6689.1	6687.2	6685.9	6666.2	6665.2	6664.5
6.000	7323.4	7321.1	7318.2	7289.4	7287.1	7285.6	7262.5	7261.3	7260.5
6.250	7957.5	7954.7	7951.2	7916.4	7913.7	7911.9	7884.7	7883.4	7882.5
6.500	8619.9	8616.5	8612.2	8570.2	8566.9	8564.8	8533.0	8531.4	8530.4
6.750	9311.1	9306.9	9301.6	9251.1	9247.2	9244.7	9207.4	9205.6	9204.4
7.000	10031	10026	10020	9959.3	9954.6	9951.7	9908.0	9905.9	9904.6
7.250	10782	10776	10768	10695	10689	10686	10635	10633	10631
7.500	11562	11555	11546	11449	11452	11448	11388	11386	11384
7.750	12374	12365	12354	12250	12243	12238	12168	12165	12163
8.000	13218	13207	13194	13071	13062	13056	12975	12971	12969
8.250	14095	14082	14066	13920	13910	13903	13809	13805	13802
8.500	15005	14990	14971	14799	14787	14779	14669	14665	14661
8.750	15950	15933	15911	15708	15693	15684	15557	15552	15548
9.000	16932	16911	16885	16647	16630	16620	16473	16466	16462
9.250	17950	17926	17895	17618	17598	17585	17416	17409	17404
9.500	19007	18979	18943	18620	18597	18582	18387	18379	18373
9.750	20104	20071	20030	19655	19628	19611	19386	19377	19371
10.000	21243	21205	21157	20723	20692	20672	20414	20403	20396
10.250	22426	22382	22325	21825	21789	21767	21471	21458	21450
10.500	23654	23603	23538	22962	22921	22895	22556	22542	22533
10.750	24931	24872	24797	24134	24088	24058	23672	23656	23646
11.000	26257	26190	26104	25344	25290	25257	24817	24799	24787
11.250	27636	27559	27460	26592	26531	26492	25992	25972	25959
11.500	29070	28982	28870	27878	27809	27766	27199	27176	27161
11.750	30563	30462	30334	29205	29126	29077	28436	28411	28394
12.000	32116	32001	31856	30574	30485	30429	29706	29677	29659
12.500	35417	35270	35084	33444	33330	33259	32343	32306	32283
13.000	39003	38817	38581	36502	36357	36267	35114	35068	35039
13.500	42913	42673	42375	39762	39581	39467	38025	37968	37932
14.000	47244	46921	46523	43241	43015	42874	41082	41012	40968
14.500				46958	46679	46505	44291	44206	44152
15.000				50934	50591	50378	47661	47557	47490
15.500				55192	54774	54513	51202	51075	50994
16.000				59759	59251	58935	54923	54769	54671
16.500				64701	64060	63671	58835	58649	58531
17.000					69288	68792	62950	62728	62586
17.500							67282	67017	66848
18.000							71846	71530	71330
18.500							76653	76282	76046
19.000							81725	81289	81011
19.500							87078	86568	86244
20.000							92734	92140	91761
20.500							98728	98025	97584
21.000							105134	104282	103752
21.500								110983	110340

TABLE E.2 "b" Values For Reynolds Number Factor F, Determination—Flange Taps

$$F_r = 1 + \frac{b}{\sqrt{h_w p_f}}$$

Pipe Sizes—Nominal and Published Inside Diameters, Inches

Orifice Diameter Inches	2			3				4	
	1.689	1.939	2.067	2.300	2.626	2.900	3.068	3.152	3.438
.250	.0879	.0911	.0926	.0950	.0979	.0999	.1010	.1014	.1030
.375	.0677	.0709	.0726	.0755	.0792	.0820	.0836	.0844	.0867
.500	.0562	.0576	.0588	.0612	.0648	.0677	.0695	.0703	.0730
.625	.0520	.0505	.0506	.0516	.0541	.0566	.0583	.0591	.0618
.750	.0536	.0485	.0471	.0462	.0470	.0486	.0498	.0504	.0528
.875	.0595	.0506	.0478	.0445	.0429	.0433	.0438	.0442	.0460
1.000	.0677	.0559	.0515	.0458	.0416	.0403	.0402	.0403	.0411
1.125	.0762	.0630	.0574	.0495	.0427	.0396	.0386	.0383	.0380
1.250	.0824	.0707	.0646	.0550	.0456	.0408	.0388	.0381	.0365
1.375		.0772	.0715	.0614	.0501	.0435	.0406	.0394	.0365
1.500			.0773	.0679	.0554	.0474	.0436	.0420	.0378
1.625				.0735	.0613	.0522	.0477	.0457	.0402
1.750					.0669	.0575	.0524	.0500	.0434
1.875					.0717	.0628	.0574	.0549	.0473
2.000						.0676	.0624	.0598	.0517
2.125						.0715	.0669	.0644	.0563
2.250							.0706	.0685	.0607
2.375									.0648
2.500									.0683

Orifice Diameter Inches	4		6				8		
	3.826	4.026	4.897	5.189	5.761	6.065	7.625	7.981	8.071
.250	.1047	.1054							
.375	.0894	.0907							
.500	.0763	.0779	.0836	.0852	.0880	.0892			
.625	.0653	.0670	.0734	.0753	.0785	.0801			
.750	.0561	.0578	.0645	.0665	.0701	.0718			
.875	.0487	.0502	.0567	.0587	.0625	.0643	.0723	.0738	.0742
1.000	.0430	.0442	.0500	.0520	.0557	.0576	.0660	.0676	.0680
1.125	.0388	.0396	.0444	.0462	.0498	.0517	.0602	.0619	.0623
1.250	.0361	.0364	.0399	.0414	.0447	.0464	.0549	.0566	.0571
1.375	.0347	.0344	.0363	.0375	.0403	.0419	.0501	.0518	.0523
1.500	.0345	.0336	.0336	.0344	.0367	.0381	.0457	.0474	.0479
1.625	.0354	.0338	.0318	.0322	.0337	.0348	.0418	.0435	.0439
1.750	.0372	.0350	.0307	.0306	.0314	.0322	.0383	.0399	.0403
1.875	.0398	.0370	.0305	.0298	.0298	.0303	.0353	.0366	.0371
2.000	.0430	.0395	.0308	.0296	.0296	.0287	.0327	.0340	.0343
2.125	.0467	.0427	.0318	.0300	.0281	.0278	.0304	.0315	.0318
2.250	.0507	.0462	.0334	.0310	.0281	.0274	.0286	.0295	.0297
2.375	.0546	.0501	.0354	.0324	.0286	.0274	.0271	.0278	.0280
2.500	.0589	.0540	.0378	.0342	.0295	.0279	.0259	.0264	.0265
2.625	.0626	.0579	.0406	.0365	.0308	.0287	.0251	.0253	.0254
2.750	.0659	.0615	.0436	.0391	.0324	.0300	.0246	.0245	.0245
2.875		.0647	.0468	.0418	.0343	.0314	.0244	.0240	.0240
3.000		.0673	.0500	.0448	.0366	.0332	.0245	.0238	.0237
3.125			.0533	.0479	.0389	.0353	.0248	.0239	.0237
3.250			.0564	.0510	.0416	.0375	.0254	.0242	.0240
3.375			.0594	.0541	.0443	.0400	.0263	.0248	.0244
3.500			.0620	.0569	.0472	.0426	.0273	.0255	.0251
3.625			.0643	.0597	.0500	.0452	.0286	.0265	.0260
3.750				.0621	.0527	.0479	.0300	.0276	.0271
3.875				.0640	.0553	.0505	.0316	.0289	.0283
4.000					.0578	.0531	.0334	.0304	.0297
4.250					.0620	.0679	.0372	.0338	.0330
4.500						.0618	.0414	.0375	.0366
4.750							.0457	.0416	.0405
5.000							.0500	.0457	.0446
5.250							.0539	.0497	.0487
5.500							.0574	.0535	.0524
5.750								.0569	.0559
6.000									.0588

"b" Values For Reynolds Number Factor F, Determination—Flange Taps

$$F_r = 1 + \frac{b}{\sqrt{h_w\, p_f}}$$

Pipe Sizes — Nominal and Published Inside Diameters, Inches

Orifice Diameter Inches	20			24			30		
	18.814	19.000	19.250	22.626	23.000	23.250	28.628	29.000	29.250
2.000	.0667	.0671	.0676						
2.125	.0640	.0644	.0649						
2.250	.0614	.0618	.0622						
2.375	.0588	.0592	.0597	.0659	.0665	.0669			
2.500	.0563	.0568	.0573	.0636	.0642	.0646			
2.625	.0540	.0544	.0549	.0614	.0620	.0624			
2.750	.0517	.0521	.0526	.0592	.0599	.0603			
2.875	.0494	.0499	.0504	.0571	.0578	.0582	.0662		
3.000	.0473	.0477	.0483	.0551	.0557	.0562	.0644	.0649	.0652
3.125	.0452	.0457	.0462	.0531	.0538	.0542	.0626	.0631	.0634
3.250	.0433	.0437	.0442	.0511	.0518	.0523	.0608	.0613	.0616
3.375	.0414	.0418	.0423	.0493	.0500	.0504	.0590	.0596	.0599
3.500	.0395	.0399	.0405	.0474	.0481	.0486	.0574	.0579	.0582
3.625	.0378	.0382	.0387	.0457	.0464	.0468	.0557	.0562	.0566
3.750	.0361	.0365	.0370	.0440	.0447	.0451	.0541	.0546	.0550
3.875	.0345	.0349	.0354	.0423	.0430	.0435	.0525	.0530	.0534
4.000	.0329	.0333	.0339	.0407	.0414	.0419	.0509	.0515	.0518
4.250	.0301	.0304	.0310	.0376	.0384	.0388	.0479	.0485	.0488
4.500	.0275	.0279	.0283	.0348	.0355	.0360	.0450	.0456	.0460
4.750	.0252	.0256	.0260	.0322	.0328	.0333	.0423	.0429	.0433
5.000	.0232	.0235	.0239	.0297	.0304	.0308	.0397	.0403	.0407
5.250	.0214	.0217	.0220	.0275	.0281	.0285	.0373	.0378	.0382
5.500	.0199	.0201	.0204	.0254	.0260	.0264	.0349	.0355	.0359
5.750	.0186	.0188	.0191	.0236	.0241	.0245	.0327	.0333	.0337
6.000	.0176	.0177	.0179	.0219	.0224	.0228	.0306	.0312	.0316
6.250	.0167	.0168	.0170	.0204	.0208	.0212	.0287	.0292	.0296
6.500	.0161	.0162	.0163	.0191	.0195	.0198	.0269	.0274	.0277
6.750	.0157	.0157	.0157	.0179	.0183	.0185	.0252	.0257	.0260
7.000	.0155	.0155	.0154	.0169	.0172	.0174	.0236	.0240	.0244
7.250	.0155	.0154	.0153	.0161	.0163	.0165	.0221	.0226	.0229
7.500	.0157	.0155	.0154	.0154	.0156	.0157	.0208	.0212	.0215
7.750	.0160	.0158	.0156	.0148	.0150	.0151	.0195	.0199	.0202
8.000	.0166	.0163	.0160	.0144	.0145	.0146	.0184	.0187	.0190
8.250	.0172	.0169	.0165	.0142	.0142	.0142	.0174	.0177	.0179
8.500	.0180	.0177	.0172	.0141	.0140	.0140	.0164	.0168	.0170
8.750	.0190	.0186	.0180	.0141	.0140	.0139	.0156	.0159	.0161
9.000	.0201	.0196	.0190	.0143	.0141	.0140	.0149	.0152	.0153
9.250	.0213	.0208	.0201	.0146	.0143	.0141	.0143	.0145	.0146
9.500	.0226	.0220	.0213	.0150	.0146	.0144	.0138	.0139	.0141
9.750	.0240	.0234	.0226	.0155	.0150	.0147	.0133	.0135	.0136
10.000	.0256	.0249	.0240	.0161	.0155	.0152	.0130	.0131	.0132
10.250	.0271	.0264	.0255	.0168	.0162	.0158	.0128	.0128	.0128
10.500	.0288	.0280	.0270	.0176	.0169	.0164	.0126	.0126	.0126
10.750	.0305	.0297	.0286	.0185	.0176	.0172	.0125	.0125	.0125
11.000	.0322	.0314	.0303	.0194	.0186	.0181	.0125	.0124	.0124
11.250	.0340	.0332	.0320	.0205	.0196	.0190	.0126	.0125	.0124
11.500	.0358	.0349	.0338	.0216	.0207	.0200	.0128	.0126	.0125
11.750	.0376	.0367	.0355	.0228	.0218	.0211	.0130	.0128	.0127
12.000	.0394	.0365	.0373	.0241	.0230	.0223	.0134	.0131	.0129
12.500	.0429	.0420	.0406	.0267	.0255	.0248	.0142	.0138	.0136
13.000	.0463	.0454	.0442	.0296	.0282	.0274	.0153	.0148	.0145
13.500	.0494	.0485	.0474	.0326	.0311	.0302	.0166	.0160	.0157
14.000	.0520	.0512	.0502	.0356	.0341	.0331	.0182	.0175	.0171
14.500				.0386	.0370	.0360	.0199	.0192	.0187
15.000				.0415	.0400	.0390	.0218	.0209	.0204
15.500				.0443	.0426	.0418	.0239	.0230	.0224
16.000				.0470	.0455	.0446	.0260	.0250	.0244
16.500				.0494	.0480	.0471	.0283	.0273	.0266
17.000					.0503	.0494	.0307	.0296	.0288
17.500							.0331	.0319	.0312
18.000							.0355	.0343	.0335
18.500							.0379	.0366	.0358
19.000							.0402	.0390	.0382
19.500							.0424	.0412	.0404
20.000							.0446	.0434	.0426
20.500							.0466	.0455	.0448
21.000							.0485	.0475	.0467
21.500								.0492	.0485

"b" Values For Reynolds Number Factor F, Determination—Flange Taps

$$F_r = 1 + \frac{b}{\sqrt{h_w\, p_f}}$$

Pipe Sizes—Nominal and Published Inside Diameters, Inches

Orifice Diameter. Inches	10			12			16		
	9.564	10.020	10.136	11.376	11.938	12.090	14.688	15.000	15.250
1.000	.0738								
1.125	.0685	.0701	.0705						
1.250	.0635	.0652	.0656	.0698	.0714	.0718			
1.375	.0588	.0606	.0610	.0654	.0671	.0676			
1.500	.0545	.0563	.0568	.0612	.0631	.0635	.0706	.0713	
1.625	.0504	.0523	.0527	.0573	.0592	.0597	.0670	.0678	.0684
1.750	.0467	.0485	.0490	.0536	.0555	.0560	.0636	.0644	.0650
1.875	.0433	.0451	.0455	.0501	.0521	.0526	.0604	.0612	.0618
2.000	.0401	.0419	.0423	.0469	.0488	.0492	.0572	.0581	.0587
2.125	.0372	.0389	.0393	.0438	.0458	.0463	.0542	.0551	.0558
2.250	.0346	.0362	.0360	.0410	.0429	.0434	.0514	.0523	.0529
2.375	.0322	.0337	.0341	.0383	.0402	.0407	.0487	.0496	.0502
2.500	.0302	.0315	.0319	.0359	.0377	.0382	.0461	.0470	.0476
2.625	.0283	.0296	.0299	.0336	.0354	.0358	.0436	.0445	.0452
2.750	.0267	.0278	.0281	.0316	.0332	.0336	.0413	.0422	.0428
2.875	.0254	.0263	.0265	.0297	.0312	.0317	.0391	.0399	.0406
3.000	.0243	.0250	.0252	.0278	.0294	.0298	.0370	.0378	.0385
3.125	.0234	.0239	.0241	.0264	.0278	.0282	.0350	.0358	.0365
3.250	.0226	.0230	.0231	.0251	.0263	.0266	.0331	.0339	.0346
3.375	.0221	.0223	.0224	.0239	.0250	.0253	.0314	.0321	.0328
3.500	.0219	.0218	.0218	.0229	.0238	.0241	.0298	.0305	.0311
3.625	.0218	.0214	.0214	.0221	.0226	.0230	.0282	.0290	.0295
3.750	.0218	.0213	.0212	.0214	.0219	.0221	.0268	.0275	.0281
3.875	.0221	.0213	.0211	.0208	.0212	.0213	.0255	.0262	.0267
4.000	.0225	.0214	.0212	.0204	.0206	.0207	.0243	.0249	.0254
4.250	.0238	.0222	.0219	.0200	.0198	.0198	.0223	.0228	.0232
4.500	.0256	.0236	.0231	.0201	.0195	.0194	.0206	.0210	.0213
4.750	.0279	.0254	.0249	.0207	.0196	.0194	.0193	.0196	.0198
5.000	.0307	.0277	.0270	.0217	.0202	.0199	.0184	.0185	.0187
5.250	.0337	.0303	.0295	.0231	.0212	.0208	.0178	.0178	.0179
5.500	.0370	.0332	.0323	.0249	.0226	.0221	.0176	.0174	.0174
5.750	.0404	.0363	.0354	.0270	.0243	.0237	.0176	.0174	.0172
6.000	.0438	.0396	.0386	.0294	.0263	.0255	.0180	.0176	.0173
6.250	.0473	.0429	.0418	.0320	.0285	.0277	.0186	.0180	.0177
6.500	.0505	.0462	.0451	.0347	.0309	.0300	.0195	.0188	.0183
6.750	.0536	.0493	.0483	.0376	.0335	.0325	.0206	.0198	.0192
7.000	.0562	.0523	.0513	.0406	.0362	.0351	.0220	.0210	.0202
7.250		.0550	.0540	.0435	.0390	.0379	.0235	.0224	.0216
7.500		.0572	.0564	.0463	.0418	.0407	.0252	.0240	.0230
7.750				.0491	.0446	.0434	.0271	.0257	.0246
8.000				.0517	.0473	.0461	.0291	.0276	.0264
8.250				.0540	.0498	.0487	.0312	.0296	.0283
8.500				.0560	.0522	.0511	.0334	.0317	.0303
8.750					.0543	.0534	.0357	.0338	.0324
9.000						.0553	.0380	.0361	.0346
9.250							.0402	.0383	.0368
9.500							.0425	.0406	.0390
9.750							.0447	.0428	.0412
10.000							.0469	.0449	.0434
10.250							.0489	.0470	.0455
10.500							.0508	.0490	.0475
10.750							.0526	.0509	.0495
11.000							.0541	.0526	.0513
11.250								.0541	.0528

TABLE E.3 Y₁ Expansion Factors—Flange Taps

Static Pressure Taken from Upstream Taps

$$\beta = \frac{d}{D} \text{ Ratio}$$

$\frac{h_w}{P_{f1}}$ Ratio	.1	.2	.3	.4	.45	.50	.52	.54	.56	.58	.60	.61	.62	.63	.64	.65	.66	.67	.68	.69	.70	.71	.72	.73	.74	.75
0.0	1.0000	1.0000	1.0000	1.0000	1.0000	1.0000	1.0000	1.0000	1.0000	1.0000	1.0000	1.0000	1.0000	1.0000	1.0000	1.0000	1.0000	1.0000	1.0000	1.0000	1.0000	1.0000	1.0000	1.0000	1.0000	1.0000
0.1	.9989	.9989	.9989	.9988	.9988	.9988	.9988	.9988	.9988	.9988	.9987	.9987	.9987	.9987	.9987	.9987	.9987	.9987	.9987	.9986	.9986	.9986	.9986	.9986	.9986	.9986
0.2	.9977	.9977	.9977	.9977	.9976	.9976	.9976	.9976	.9975	.9975	.9975	.9975	.9974	.9974	.9974	.9974	.9974	.9973	.9973	.9973	.9973	.9972	.9972	.9972	.9971	.9971
0.3	.9966	.9966	.9966	.9965	.9965	.9964	.9964	.9964	.9963	.9963	.9963	.9962	.9962	.9962	.9961	.9961	.9961	.9960	.9960	.9959	.9959	.9958	.9958	.9958	.9957	.9957
0.4	.9954	.9954	.9954	.9953	.9953	.9952	.9952	.9952	.9951	.9951	.9950	.9949	.9949	.9949	.9948	.9948	.9948	.9947	.9946	.9946	.9945	.9945	.9944	.9943	.9943	.9942
0.5	.9943	.9943	.9943	.9942	.9941	.9940	.9940	.9939	.9939	.9938	.9938	.9937	.9936	.9936	.9935	.9935	.9934	.9934	.9933	.9933	.9932	.9931	.9931	.9930	.9929	.9928
0.6	.9932	.9932	.9931	.9930	.9929	.9928	.9927	.9927	.9926	.9925	.9924	.9924	.9923	.9923	.9922	.9921	.9921	.9920	.9919	.9918	.9918	.9917	.9916	.9915	.9914	.9913
0.7	.9920	.9920	.9920	.9919	.9918	.9916	.9915	.9915	.9914	.9913	.9912	.9911	.9910	.9910	.9909	.9908	.9907	.9907	.9906	.9905	.9904	.9903	.9902	.9901	.9900	.9899
0.8	.9909	.9909	.9908	.9907	.9906	.9904	.9903	.9903	.9901	.9900	.9899	.9898	.9898	.9897	.9897	.9896	.9895	.9894	.9893	.9892	.9891	.9890	.9889	.9887	.9886	.9884
0.9	.9898	.9897	.9897	.9895	.9894	.9892	.9891	.9890	.9889	.9888	.9887	.9886	.9885	.9885	.9884	.9883	.9882	.9881	.9880	.9879	.9878	.9877	.9875	.9874	.9873	.9870
1.0	.9886	.9886	.9885	.9884	.9882	.9880	.9879	.9878	.9877	.9875	.9874	.9873	.9872	.9871	.9870	.9870	.9869	.9868	.9867	.9865	.9864	.9863	.9861	.9860	.9859	.9857
1.1	.9875	.9875	.9874	.9872	.9870	.9868	.9867	.9866	.9864	.9863	.9861	.9860	.9859	.9858	.9857	.9856	.9854	.9853	.9852	.9851	.9849	.9848	.9846	.9844	.9843	.9841
1.2	.9863	.9863	.9862	.9860	.9859	.9856	.9855	.9853	.9852	.9850	.9848	.9847	.9846	.9845	.9844	.9843	.9841	.9840	.9838	.9837	.9835	.9834	.9832	.9830	.9828	.9826
1.3	.9852	.9852	.9851	.9849	.9847	.9844	.9843	.9841	.9840	.9838	.9836	.9835	.9833	.9832	.9831	.9829	.9828	.9827	.9825	.9823	.9822	.9820	.9818	.9816	.9814	.9812
1.4	.9841	.9840	.9840	.9837	.9835	.9832	.9831	.9829	.9827	.9825	.9823	.9822	.9821	.9819	.9818	.9816	.9815	.9813	.9812	.9810	.9808	.9806	.9804	.9802	.9800	.9798
1.5	.9829	.9829	.9828	.9826	.9823	.9820	.9819	.9817	.9815	.9813	.9810	.9809	.9808	.9806	.9805	.9803	.9802	.9800	.9798	.9796	.9794	.9792	.9790	.9788	.9786	.9783
1.6	.9818	.9818	.9817	.9814	.9811	.9808	.9806	.9805	.9803	.9800	.9798	.9796	.9795	.9793	.9792	.9790	.9788	.9787	.9785	.9783	.9781	.9778	.9776	.9774	.9771	.9769
1.7	.9806	.9806	.9805	.9802	.9800	.9796	.9794	.9792	.9790	.9788	.9785	.9784	.9782	.9780	.9779	.9777	.9775	.9773	.9771	.9769	.9767	.9764	.9762	.9760	.9757	.9754
1.8	.9795	.9795	.9794	.9791	.9788	.9784	.9782	.9780	.9778	.9775	.9772	.9771	.9769	.9768	.9766	.9764	.9762	.9760	.9758	.9755	.9753	.9751	.9748	.9745	.9743	.9740
1.9	.9784	.9783	.9782	.9779	.9776	.9772	.9770	.9768	.9766	.9763	.9760	.9758	.9756	.9755	.9753	.9751	.9749	.9747	.9744	.9742	.9739	.9737	.9734	.9731	.9728	.9725
2.0	.9772	.9772	.9771	.9767	.9764	.9760	.9758	.9756	.9753	.9750	.9747	.9745	.9744	.9742	.9740	.9738	.9735	.9733	.9731	.9728	.9726	.9723	.9720	.9717	.9714	.9711
2.1	.9761	.9761	.9759	.9756	.9753	.9748	.9746	.9744	.9741	.9738	.9734	.9733	.9731	.9729	.9727	.9725	.9722	.9720	.9717	.9715	.9712	.9709	.9706	.9703	.9700	.9696
2.2	.9750	.9749	.9748	.9744	.9741	.9736	.9734	.9731	.9729	.9725	.9722	.9720	.9718	.9716	.9714	.9711	.9709	.9706	.9704	.9701	.9698	.9695	.9692	.9689	.9685	.9682
2.3	.9738	.9738	.9736	.9732	.9729	.9724	.9722	.9719	.9716	.9713	.9709	.9707	.9705	.9703	.9701	.9698	.9696	.9693	.9690	.9688	.9685	.9681	.9678	.9675	.9671	.9667
2.4	.9727	.9726	.9725	.9721	.9717	.9712	.9710	.9707	.9704	.9700	.9697	.9694	.9692	.9690	.9688	.9685	.9683	.9680	.9677	.9674	.9671	.9668	.9664	.9661	.9657	.9653
2.5	.9715	.9715	.9713	.9709	.9705	.9700	.9698	.9695	.9692	.9688	.9684	.9682	.9680	.9677	.9675	.9672	.9669	.9666	.9663	.9660	.9657	.9654	.9650	.9647	.9643	.9639
2.6	.9704	.9704	.9702	.9698	.9694	.9688	.9686	.9683	.9679	.9675	.9671	.9669	.9667	.9664	.9662	.9659	.9656	.9653	.9650	.9647	.9643	.9640	.9636	.9632	.9628	.9624
2.7	.9693	.9692	.9691	.9686	.9682	.9676	.9673	.9670	.9667	.9663	.9659	.9656	.9654	.9651	.9649	.9646	.9643	.9640	.9637	.9633	.9630	.9626	.9622	.9618	.9614	.9610
2.8	.9681	.9681	.9679	.9674	.9670	.9664	.9661	.9658	.9654	.9650	.9646	.9644	.9641	.9638	.9636	.9633	.9630	.9626	.9623	.9620	.9616	.9612	.9608	.9604	.9600	.9595
2.9	.9670	.9669	.9668	.9663	.9658	.9652	.9649	.9646	.9642	.9638	.9633	.9631	.9628	.9625	.9623	.9620	.9616	.9613	.9610	.9606	.9602	.9598	.9594	.9590	.9585	.9581
3.0	.9658	.9658	.9656	.9651	.9647	.9640	.9637	.9634	.9630	.9626	.9621	.9618	.9615	.9613	.9610	.9606	.9603	.9600	.9596	.9592	.9588	.9584	.9580	.9576	.9571	.9566
3.1	.9647	.9647	.9645	.9639	.9635	.9628	.9625	.9622	.9617	.9613	.9608	.9605	.9603	.9600	.9597	.9593	.9590	.9586	.9583	.9579	.9575	.9571	.9566	.9562	.9557	.9552
3.2	.9636	.9635	.9633	.9628	.9623	.9616	.9613	.9609	.9605	.9601	.9595	.9593	.9590	.9587	.9584	.9580	.9577	.9573	.9569	.9565	.9561	.9557	.9552	.9547	.9542	.9537
3.3	.9624	.9624	.9622	.9616	.9611	.9604	.9601	.9597	.9593	.9588	.9583	.9580	.9577	.9574	.9571	.9567	.9564	.9560	.9556	.9552	.9547	.9543	.9538	.9533	.9528	.9523
3.4	.9613	.9612	.9610	.9604	.9599	.9592	.9589	.9585	.9580	.9576	.9570	.9567	.9564	.9561	.9558	.9554	.9550	.9546	.9542	.9538	.9534	.9529	.9524	.9519	.9514	.9508
3.5	.9602	.9601	.9599	.9593	.9588	.9580	.9577	.9573	.9568	.9563	.9558	.9554	.9551	.9548	.9545	.9541	.9537	.9533	.9529	.9524	.9520	.9515	.9510	.9505	.9500	.9494
3.6	.9590	.9590	.9587	.9581	.9576	.9568	.9565	.9560	.9556	.9551	.9545	.9542	.9538	.9535	.9532	.9528	.9524	.9520	.9515	.9511	.9506	.9501	.9496	.9491	.9485	.9480
3.7	.9579	.9578	.9576	.9570	.9564	.9556	.9553	.9548	.9543	.9538	.9532	.9529	.9526	.9522	.9518	.9515	.9511	.9506	.9502	.9497	.9492	.9487	.9482	.9477	.9471	.9465
3.8	.9567	.9567	.9564	.9558	.9552	.9544	.9540	.9536	.9531	.9526	.9520	.9516	.9513	.9509	.9505	.9502	.9497	.9493	.9488	.9484	.9479	.9474	.9468	.9463	.9457	.9451
3.9	.9556	.9555	.9553	.9546	.9540	.9532	.9528	.9524	.9519	.9513	.9507	.9504	.9500	.9496	.9492	.9488	.9484	.9480	.9475	.9470	.9465	.9460	.9454	.9448	.9442	.9436
4.0	.9545	.9544	.9542	.9535	.9529	.9520	.9516	.9512	.9506	.9501	.9494	.9491	.9487	.9483	.9479	.9475	.9471	.9465	.9462	.9457	.9451	.9446	.9440	.9434	.9428	.9422

TABLE E.4 Y_2 Expansion Factors—Flange Taps

Static Pressure Taken from Downstream Taps

$\frac{h_w}{P_{f2}}$ Ratio	$\beta = \frac{d}{D}$ Ratio																									
	.1	.2	.3	.4	.45	.50	.52	.54	.56	.58	.60	.61	.62	.63	.64	.65	.66	.67	.68	.69	.70	.71	.72	.73	.74	.75
0.0	1.0000	1.0000	1.0000	1.0000	1.0000	1.0000	1.0000	1.0000	1.0000	1.0000	1.0000	1.0000	1.0000	1.0000	1.0000	1.0000	1.0000	1.0000	1.0000	1.0000	1.0000	1.0000	1.0000	1.0000	1.0000	1.0000

Appendix F

Sample Data Reporting Forms

Figures reprinted from *The Oil and Gas Conservation Act,* and *Oil and Gas Conservation Regulations,* by permission of Energy Resources Conservation Board, Alberta, Canada.

GAS WELL DELIVERABILITY TEST - FIELD NOTES

ENERGY RESOURCES CONSERVATION BOARD
ALBERTA CANADA

PAGE 1 OF _____

WELL NAME _____

UNIQUE IDENTIFIER

S	LE	LS	SC	TWP	RG	W	M	E	S

FIELD OR AREA _____ POOL OR ZONE _____

PERF/OPEN HOLE INTERVAL (KB)_____ m   PRODUCING THROUGH: TUBING ☐   ANNULUS ☐

WELL BLOWN FOR _____ min   SPRAY: WATER/CONDENSATE CLEAR IN_____ min

DATE SHUT-IN [YR | MO | DAY] [TIME]   TOTAL SHUT-IN TIME _____ h

SHUT-IN NO 1 (INITIAL)

| DATE | | | TIME | CUM. SHUT-IN TIME h | WELLHEAD PRESSURE (Gauge) kPa | | WELLHEAD TEMPERATURE °C |
YR	MO	DAY			TUBING	CASING	

REMARKS

FLOW NO.1   WELL OPENED AT [YR | MO | DAY] [TIME]

| DATE | | | TIME | CUM. FLOW TIME h | WELLHEAD PRESSURE (Gauge) kPa | | WELLHEAD TEMPERATURE °C | METER OR PROVER DATA | | |
YR	MO	DAY			TUBING	CASING		STATIC PRESSURE (Gauge) kPa	DIFFERENTIAL mm H₂O	TEMPERATURE °C

METER RUN OR PROVER SIZE _____ mm   ORIFICE SIZE _____ mm

SEPARATOR CONDITIONS: (HP SEP.) P_{ga} _____ kPa, T _____ °C   (LP SEP.) P_{ga} _____ kPa, T _____ °C

CONDENSATE PRODUCTION RATE _____ m³/d (HP,LP,ST)   TOTAL _____ m³

WATER PRODUCTION RATE _____ m³/d   TOTAL _____ m³

FINAL FLOWING WELLHEAD PRESSURE (Gauge) TUBING _____ kPa, CASING _____ kPa

WELL SHUT-IN AT [YR | MO | DAY] [TIME]   TOTAL FLOW TIME _____ h

NOTE FLOWING WELLHEAD PRESSURES AND TEMPERATURES MUST BE OBTAINED UPSTREAM OF ANY CHOKING DEVICE

EG-29A-79-07

Figure F.1
Sample of Gas Well Deliverability Test-Field Notes.

GAS WELL DELIVERABILITY TEST - FIELD NOTES

ENERGY RESOURCES CONSERVATION BOARD
ALBERTA, CANADA

PAGE _____ OF _____

SHUT-IN NO. _____ (INTERMEDIATE)

DATE			TIME	CUM SHUT-IN TIME h	WELLHEAD PRESSURE (Gauge) kPa		WELLHEAD TEMPERATURE °C
YR	MO	DAY			TUBING	CASING	
			:				
			:				
			:				
			:				
			:				
			:				
			:				
			:				
			:				
			:				
			:				
			:				

REMARKS

FLOW NO _____   WELL OPENED AT [YR MO DAY] [TIME]

DATE			TIME	CUM FLOW TIME h	WELLHEAD PRESSURE (Gauge) kPa		WELLHEAD TEMPERATURE °C	METER OR PROVER DATA		
YR	MO	DAY			TUBING	CASING		STATIC PRESSURE (Gauge) kPa	DIFFERENTIAL mm H2O	TEMPERATURE °C
			:							
			:							
			:							
			:							
			:							
			:							
			:							
			:							
			:							
			:							
			:							
			:							

METER RUN OR PROVER SIZE _____ mm   ORIFICE SIZE _____ mm

SEPARATOR CONDITIONS (HP SEP) P_{ga} _____ kPa, T _____ °C   (LP SEP) P_{ga} _____ kPa, T _____ °C

CONDENSATE PRODUCTION RATE _____ m³/d (HP,LP,ST)   TOTAL _____ m³

WATER PRODUCTION RATE _____ m³/d   TOTAL _____ m³

FINAL FLOWING WELLHEAD PRESSURE (Gauge) TUBING _____ kPa   CASING _____ kPa

WELL SHUT-IN AT [YR MO DAY] [TIME]   TOTAL FLOW TIME _____ h

EG-29B-79-07

Figure F.2
Sample of Gas Well Deliverability Test-Field Notes.

GAS WELL DELIVERABILITY TEST – FIELD NOTES

ENERGY RESOURCES CONSERVATION BOARD
ALBERTA, CANADA

PAGE _____ OF _____

CONTINUATION OF FLOW NO _____ TO STABILIZATION

| DATE | | | TIME | CUM. FLOW TIME h | WELLHEAD PRESSURE (Gauge) kPa | | WELLHEAD TEMPERATURE °C | METER OR PROVER DATA | | |
YR	MO	DAY			TUBING	CASING		STATIC PRESSURE (Gauge) kPa	DIFFERENTIAL mm H2O	TEMPERATURE °C
			:							
			:							
			:							
			:							
			:							
			:							
			:							
			:							
			:							
			:							
			:							
			:							
			:							
			:							
			:							
			:							
			:							
			:							
			:							
			:							
			:							
			:							
			:							

METER RUN OR PROVER SIZE _____ mm   ORIFICE SIZE _____ mm

SEPARATOR CONDITIONS: (HP SEP.) P_{ga} _____ kPa, T _____ °C   (LP SEP.) P_{ga} _____ kPa, T _____ °C

CONDENSATE PRODUCTION RATE _____ m^3/d (HP, LP, ST)   TOTAL _____ m^3

WATER PRODUCTION RATE _____ m^3/d TOTAL _____ m^3

FINAL FLOWING WELLHEAD PRESSURE (Gauge): TUBING _____ kPa CASING _____ kPa

WELL SHUT-IN AT | | | | (YR MO DAY) | : | (TIME)   TOTAL FLOW TIME _____ h

FINAL SHUT-IN WELLHEAD PRESSURE (Gauge): TUBING _____ kPa CASING _____ kPa

DURATION OF FINAL SHUT-IN _____ h   TESTED BY (CO.) _____

EG-29C-79-07

Figure F.3
Sample of Gas Well Deliverability Test-Field Notes.

GAS WELL DELIVERABILITY TEST SUMMARY
(ALL PRESSURES ABSOLUTE)

ENERGY RESOURCES CONSERVATION BOARD
ALBERTA, CANADA

WELL NAME_____ UNIQUE WELL IDENTIFIER | S | LE | LS | SC | TWP | RG | W | M | E | S |

FIELD OR AREA _____ POOL OR ZONE _____

PERF / OPEN HOLE INTERVAL _____ m (KB) ELEVATION (GL) _____ (KB) _____ m

CASING ID_____mm TUBING ID_____mm OD_____mm PACKER_____m(KB) RES. TEMP._____°C

RESERVOIR GAS PROPERTIES: G_____ P_c_____ kPa T_c_____ K MOLE FRACTION: N_2____ CO_2____ H_2S____

LICENCEE: _____ OPERATOR (CO.) _____

TYPE OF TEST _____ FINAL DATE OF TEST | | | |
YR   MO   DAY

PRODUCTION DATA

RATE NO.	DURATION h	GAS PRODUCTION $10^3 m^3/d$ (st)	CONDENSATE PRODUCTION m^3/d	COND/GAS RATIO 10^{-3}	GAS EQUIVALENT OF CONDENSATE $10^3 m^3$ (st)	TOTAL PRODUCTION RATE $10^3 m^3/d$ (st)	WATER PRODUCTION m^3/d
EXTENDED RATE							

GAS PRODUCED THROUGH: TUBING ☐ CASING ☐ TO PIPELINE ☐ VENT ☐ FLARE ☐

FLARE STACK HEIGHT _____ m DIAMETER _____ mm

TOTAL VOLUME OF GAS PRODUCED DURING CLEANUP AND TEST _____ $10^3 m^3$ (st)

EQUIPMENT LIST: LINE HEATER ☐ LP SEPARATOR ☐ HP SEPARATOR ☐ CRIT. FLOW PROVER ☐

ORIFICE METER ☐ LIQUID STORAGE TANK ☐ OTHER _____

DELIVERABILITY TEST CALCULATIONS - SIMPLIFIED ANALYSIS
(BASE CONDITIONS = 101.325 kPa and 15°C) (NOTE: q IMPLIES q_{st})

	DURATION h	SANDFACE PRESSURE p kPa	CALC	MEAS.	p^2 $10^6 kPa^2$	Δp^2 $10^6 kPa^2$	FLOW RATE q $10^3 \frac{m^3}{d}$	RESULTS
INITIAL SHUT-IN								
FLOW 1								$q = C(\overline{P}_R^2 - P_{wf}^2)^n$
SHUT-IN								
FLOW 2								inverse slope n _____
SHUT-IN								$\overline{P}_R$ = _____ kPa
FLOW 3								$C = \dfrac{q}{(\overline{P}_R^2 - P_{wf}^2)^n}$
SHUT-IN								
FLOW 4								= _____
EXTENDED FLOW								AOF =
FINAL SHUT-IN								_____ m^3/d (st)

REMARKS: _____

EG-32-79-07

Figure F.4

Sample of Gas Well Deliverability Test Summary.

GAS WELL DELIVERABILITY TEST CALCULATION

(ALL PRESSURES ABSOLUTE)

ENERGY RESOURCES CONSERVATION BOARD
ALBERTA, CANADA

WELL NAME_____

UNIQUE IDENTIFIER | S | LE | LS | SC | TWP | RG | W | M E | S |

POOL OR ZONE_____ FINAL DATE OF TEST

YR MO DAY

DELIVERABILITY TEST CALCULATIONS — LIT (ψ) ANALYSIS

(BASE CONDITIONS = 101.325 kPa and 15°C) (NOTE: q IMPLIES q_{st})

	DURATION	SANDFACE PRESSURE P	ψ	$\Delta\psi$	FLOW RATE q	$\Delta\psi/q$	q^2	$\Delta\psi - bq^2$
	h	kPa	$10^3 \frac{MPa^2}{Pa \cdot s}$	$10^3 \frac{MPa^2}{Pa \cdot s}$	$10^3 \frac{m^3}{d}$	$\frac{MPa^2}{(Pa \cdot s)(m^3 \cdot d^{-1})^2}$	$10^9 \left(\frac{m^3}{d}\right)^2$	$10^3 \frac{MPa^2}{Pa \cdot s}$
INITIAL SHUT-IN								
FLOW 1								
SHUT-IN								
FLOW 2								
SHUT-IN								
FLOW 3								
SHUT-IN								
FLOW 4								
TOTAL = Σ								
EXTENDED FLOW								
FINAL SHUT-IN								

DISCARDED POINT_____

$N =$ _____ $\bar{\psi}_R =$ _____ $\frac{MPa^2}{Pa \cdot s}$

$a, a_t = \dfrac{\Sigma \frac{\Delta\psi}{q} \Sigma q^2 - \Sigma q \Sigma \Delta\psi}{N \Sigma q^2 - \Sigma q \Sigma q} =$ _____ $\frac{MPa^2}{(Pa \cdot s)(m^3 \cdot d^{-1})}$

$b = \dfrac{N \Sigma \Delta\psi - \Sigma q \Sigma \frac{\Delta\psi}{q}}{N \Sigma q^2 - \Sigma q \Sigma q} =$ _____ $\frac{MPa^2}{(Pa \cdot s)(m^3 \cdot d^{-1})}$

(EXTENDED FLOW) $\Delta\psi =$ _____

$b =$ _____ $q =$ _____

$a = \dfrac{\Delta\psi - bq^2}{q} =$ _____

RESULTS

TRANSIENT FLOW: $\bar{\psi}_R - \psi_{wf} = a_t q + bq^2$

_____ $- \psi_{wf} =$ _____ $q +$ _____ q^2

STABILIZED FLOW: $\bar{\psi}_R - \psi_{wf} = aq + bq^2$

_____ $- \psi_{wf} =$ _____ $q +$ _____ q^2

DELIVERABILITY:

$$q = \frac{1}{2b}\left[-a + \sqrt{a^2 + 4b(\bar{\psi}_R - \psi_{wf})}\right]$$

FOR $\psi_{wf} = 0$, $q = $ AOF = _____ $m^3/d(st)$

NOTE:

Although the LIT (ψ) deliverability relationship is derived by calculation, both the equation and the data points should be plotted using the simplified analysis technique. Data points which are significantly off the line representing the equation should be discarded, and the deliverability relationship then recalculated.

EG-33-80-06

Figure F.5
Sample of Gas Well Deliverability Test Calculation.

ENERGY RESOURCES CONSERVATION BOARD
PROVINCE OF ALBERTA

GAS WELL DELIVERABILITY TEST CALCULATIONS - FLOW RATES

(BASE CONDITIONS = 101 325 kPa and 15 °C)

CRITICAL FLOW PROVER

$q = {}^{\circ}C \quad P \quad F_{tf} \quad F_g \quad F_{pv} \quad (m^3 \cdot d^{-1}_{(st)})$

$F_g = \sqrt{\frac{0.6}{G}}$

RATE NO.	PROVER SIZE (mm)	ORIFICE DIAMETER (mm)	BASIC ORIFICE COEFFICIENT (C) ($m^3 \cdot d^{-1} \cdot kPa^{-1}$)	STATIC PRESSURE (P) (kPa)	FLOW TEMP FACTOR F_{tf}	RELATIVE DENSITY FACTOR F_g	SUPERCOMP FACTOR F_{cv}	FLOW RATE q ($10^3 m^3 \cdot d^{-1}_{(st)}$)
1								
2								
3								
4								
5								

(* SEE REVERSE SIDE)

ORIFICE METER

$*q = 6.762 \times 10^{-1} C' \sqrt{h_w \ P_f} \quad (m^3 \cdot d^{-1}_{(st)})$

$C' = F_b \quad F_{pb} \quad F_{tb} \quad F_g \quad F_{tf} \quad F_r \quad Y \quad F_{pv} \quad F_m$

$F_g = \sqrt{\frac{G}{G}}$

$F_{pb} = 1.0055$

$F_{tb} = 1.0000$

RATE NO.	STAGE	METER RUN OR LINE SIZE (inches)	ORIFICE DIAMETER (inches)	STATIC PRESSURE P_f (psia)	DIFFERENTIAL h_w (INCHES H_2O)	BASIC ORIFICE FACTOR F_b	SPECIFIC GRAVITY FACTOR F_g	FLOW TEMP FACTOR F_{tf}
1	H							
	L							
2	H							
	L							
3	H							
	L							
4	H							
	L							
5	H							
	L							

ORIFICE METER CALCULATIONS (CONTINUED)

RATE NO.	STAGE	REYNOLDS FACTOR F_r	EXPANSION FACTOR Y	SUPERCOMP FACTOR F_{pv}	MANOMETER FACTOR F_m	C' (ft^3/hr)	$\sqrt{h_w \ P_f}$	FLOW RATE q ($10^3 m^3 \cdot d^{-1}_{(st)}$)	TOTAL GAS PRODUCTION RATE ($10^3 m^3 \cdot d^{-1}_{(st)}$)
1	H								
	L								
2	H								
	L								
3	H								
	L								
4	H								
	L								
5	H								
	L								

EG-34-79-08   FLANGE TAPS ☐   PIPE TAPS ☐

Figure F.6
Sample of Gas Deliverability Test Calculations-Flow Rates.

Appendix F 205

ENERGY RESOURCES CONSERVATION BOARD
ALBERTA, CANADA

MOLAL RECOMBINATION CALCULATIONS

WELL _____ UNIQUE WELL IDENTIFIER

S	LE	LS	SC	TWP	RG	W	M	E	S

FIELD_____ POOL _____ DATE SAMPLED

YR	MO	DAY

SEPARATOR CONDITIONS   HP SEP, P_{ga} _____ kPa, T_____ °C   LP SEP, P_{ga} _____ kPa, T_____ °C

SEPARATOR PRODUCTS   HP GAS _____ 10^3 m^3/d (st)   LP GAS _____ 10^3 m^3/d (st)

LIQUID: STOCK TANK _____ m^3/d HP _____ m^3/d LP _____ m^3/d

LIQUID FLOW RATE CALCULATIONS*   (STOCK TANK ☐,   HP ☐,   or   LP ☐   LIQUID)

MOLAR MASS OF LIQUID, M_L = _____ g/mol,   DENSITY OF LIQUID, ρ_L = _____ kg/m^3

LIQUID FLOW RATE (mol/d)   =   FLOW RATE (m^3/d) $\times \dfrac{\rho_L}{M_L} \times$ 1 000 = _____ mol/d

GAS - EQUIVALENT OF LIQUID   =   LIQUID FLOW RATE (mol/d) $\times$ 23.645 $\times 10^{-6}$ = _____ 10^3 m^3/d (st)
(10^3 m^3/d (st))

COMP. i	M_i g/mol	T_{ci} K	P_{ci} kPa	LIQUID MOLE FRACTION	LIQUID GAS RATE $10^3 m^3/d$(st)	HP GAS MOLE FRACTION	HP GAS GAS RATE $10^3 m^3/d$(st)	LP GAS MOLE FRACTION	LP GAS GAS RATE $10^3 m^3/d$(st)	TOTAL GAS RATE $10^3 m^3/d$(st)	x_i	$x_i M_i$	$x_i T_{ci}$	$x_i P_{ci}$
N_2	28.013	126.27	3 399.1											
CO_2	44.010	304.22	7 384.3											
H_2S	34.076	373.55	9 004.6											
C_1	16.042	190.59	4 604.3											
C_2	30.070	305.43	4 880.1											
C_3	44.097	369.83	4 249.2											
iC_4	58.124	408.15	3 648.0											
nC_4	58.124	425.19	3 796.9											
iC_5	72.151	460.44	3 381.2											
nC_5	72.151	469.66	3 368.8											
C_6	86.178	507.44	3 012.3											
**C_7^+	114.232	568.84	2 486.2											
Σ				1.000 0		1.000 0		1 000 0		1.000 0				.

RECOMBINED GAS PROPERTIES: FLOW RATE = _____ $\times 10^3 m^3/d$(st),   G = _____ /28.964 = _____ , T_c = _____ K, P_c = _____ kPa

*   THE LIQUID FLOW RATE (m^3/d) AND LIQUID DENSITY ARE TO BE MEASURED AT THE SAME CONDITIONS

**   PHYSICAL PROPERTIES OF OCTANES ARE USED FOR THE C_7^+ FRACTION

BASE CONDITIONS = 101 325 kPa and 15 °C

EG - 35 - 79 - 07

Figure F.7
Sample of Molal Recombination Calculations.

ENERGY RESOURCES CONSERVATION BOARD
ALBERTA, CANADA

SUBSURFACE PRESSURE MEASUREMENTS

COMPANY_____ WELL NAME_____

DATE OF TEST [ ] [ ] [ ]   UNIQUE WELL IDENTIFIER [S] [| | |] [| W | E |]
           YR MO DAY   LE LS SC TWP RG M S

CHART READINGS AND CALCULATIONS FOR BUILD-UP OR DRAW-DOWN TEST   PAGE____ OF____

DATE			DEPTH BELOW CASING FLANGE	TIME	DEFLECTION	CALCULATED PRESSURE (GAUGE)	CORRECTION P-PC DEVIATION	CORRECTED PRESSURE (GAUGE)	PRESSURE AT MID-POINT OF PRODUCING INTERVAL
YR	MO	DAY	m		mm	kPa	kPa	kPa	kPa

O-12A-78-09

Figure F.8
Sample of Subsurface Pressure Measurements.

ENERGY RESOURCES CONSERVATION BOARD
ALBERTA, CANADA

SUBSURFACE PRESSURE MEASUREMENTS

1. BASIC DATA   PAGE ____ OF ____

COMPANY _____   WELL NAME _____
ADDRESS _____   UNIQUE WELL IDENTIFIER S | LE | LS | SC | TWP | RG | W M | E'S
FIELD and POOL _____   STATUS OIL ☐ GAS ☐ OTHER SPECIFY
TYPE OF TEST _____   DATE OF TEST [YR|MO|DAY] TO [YR|MO|DAY]
PRODUCING INTERVAL (CF) _____ m PERF ☐ OH ☐   PRODUCING THROUGH _____ mm TUBING ☐
ELEVATION (CF) _____ m(KB) _____ m(KB) to (CF) _____ m _____ mm CASING ☐
POOL DATUM (SUB SEA) _____ m   MID POINT OF PRODUCING (MPP) INTERVAL (CF) _____ m
ELEMENT SERIAL NO. _____ RANGE (GAUGE) _____ kPa   DATUM DEPTH OF WELL (FROM CF) _____ m
CALIBRATION EQUATION _____   CLOCK RANGE _____ h DATE OF LAST CALIBRATION [YR|MO|DAY]

2. STATIC TEST

PRESSURE (GAUGE) TUBING _____ kPa GAUGE ☐ DWG ☐   SHUT IN DATE [YR|MO|DAY] DURATION @ _____ h
                 CASING _____ kPa GAUGE ☐ DWG ☐   DATE ON BOTTOM [YR|MO|DAY] @ _____ h
RUN DEPTH (FROM CF) _____ m   DATE OFF BOTTOM [YR|MO|DAY] @ _____ h
B.H. TEMP. _____ °C R.D. PRESSURE (GAUGE) _____ kPa   MPP PRESSURE (GAUGE) _____ kPa
SURFACE TEMP. _____ °C GRADIENT (GAUGE) _____ kPa/m   DATUM DEPTH PRESSURE (GAUGE) _____ kPa

3. ACOUSTIC WELL SOUNDER TEST

DEPTH OF TUBING _____ m   SHUT IN DATE [YR|MO|DAY] DURATION _____ h
AVERAGE TUBING JOINT LENGTH _____ m   CURRENT WATER CUT _____ fr.
NO. OF JOINTS OF LIQUID _____   CASING PRESSURE (GAUGE) _____ kPa
LENGTH OF GAS COLUMN _____ m   GAS COLUMN PRESSURE (GAUGE) _____ kPa
LENGTH OF LIQUID COLUMN _____ m   LIQUID COLUMN PRESSURE (GAUGE) _____ kPa
GAS RELATIVE DENSITY, G _____ GAS GRADIENT (GA) _____ kPa/m   MPP PRESSURE (GAUGE) _____ kPa
OIL DENSITY ρ_L _____ kg/m³ OIL GRADIENT (GA) _____ kPa/m   DATUM DEPTH PRESSURE (GAUGE) _____ kPa

4. BUILD-UP OR DRAWDOWN TEST (Attach Forms O-12A)

RESERVOIR / WELL PARAMETERS - POROSITY _____ fr. WELL RADIUS _____ m THICKNESS _____ m WATER SATURATION _____ fr.
B.H. TEMP. _____ °C DATE AND DESCRIPTION OF LAST WELLBORE TREATMENT [YR|MO|DAY] _____

CASING I.D. _____ mm TUBING I.D. _____ mm TUBING LINEAL MASS _____ kg/m
COMPRESSIBILITY _____ kPa⁻¹
OIL WELLS ONLY: GAS WELLS ONLY:
OIL VISCOSITY _____ mPa·s   GAS VISCOSITY _____ μPa·s
BUBBLE POINT PRESSURE (GAUGE) _____ kPa   Z at RESERVOIR CONDITIONS _____
                                             RELATIVE DENSITY G _____
$B_o = \left(\dfrac{VOL. \ AT \ RES. \ COND.}{VOL. \ AT \ ST. \ COND.} \right)$ _____
WELL PRODUCING RATE DURING PRIOR TO TEST OIL _____ m³/d GAS _____ 10³m³/d (st) WATER _____ m³/d
TOTAL PRODUCTION - DURING PRIOR TO TEST OIL _____ m³ GAS _____ 10³m³/d (st) WATER _____ m³/d

5. CHART READINGS AND CALCULATIONS FOR STATIC TEST

DATE			DEPTH BELOW CF m	TIME h	DEFLECTION mm	CALCULATED PRESSURE (GAUGE) kPa	CORRECTION P-PC DEVIATION (GAUGE) kPa	CORRECTED PRESSURE (GAUGE) kPa	GRADIENT kPa/m
YR	MO	DAY							

REMARKS _____

SURVEY COMPANY _____ TEST BY _____ COMPUTED BY _____ CHECKED BY _____

012 - 79 - 08

Figure F.9
Sample of Subsurface Pressure Measurements.

Index

Melancon Heirs #1. *See* Field test
Modified isochronal test, 17–22. *See also* Isochronal test
    analysis of, 19–22
    and conventional and isochronal tests, 19
    flow rate used in, 64
    and isochronal test, 19
    use of, 63
Molal recombination calculations, sample reporting form for, 207
Multipoint backpressure test procedures
    Interstate Oil Compact Commission rules for, 84–93
    New Mexico state regulations for, 110–118
    Oklahoma state regulations for, 106–108

Negative flow rate, hypothetical, in buildup test, 43
Net pay thickness
    as determinant of flow behavior, 24
    as determinant of performance factor, 24
New Mexico, state regulations of, 109–123
    calculation, rules of, 122–123
    general instructions, 109–110
    multipoint backpressure test procedure, 110–118
    stabilized one-point backpressure test procedure, 119–122
Non-Darcy flow, 24. *See also* Darcy's law
    coefficient in single-point test, 68
    coefficient, testing procedures to determine, 41–52
    skin, 40
Non-stabilized multipoint tests, Interstate Oil Compact Commission rules for, 87–93

Oklahoma Corporation Commission, single-point test allowed by, 12
Oklahoma, state regulations of, 102–109
    gas well potentials, method of taking, 102–103
    general instructions, 103–105

multipoint backpressure test procedures, 106–108
one-point stabilized backpressure test procedures, 105–106
reporting, 108–109
One-point stabilized backpressure test procedure. *See also* Single-point test
    Interstate Oil Compact Commission rules for, 93–97
    New Mexico state regulations for, 119–122
    Oklahoma state regulations for, 105–106
Open-orifice metering system, 68
Orifice
    coefficients, 186–197
    defined, 68
Orifice meter, use of, in tests, 68–72

Performance curve in isochronal test. *See* Isochronal curve
Performance factor
    defined, 9
    determinants of, 10
Permeability, 8, 10. *See also* Skin
    in analytical solution of diffusivity equation for a single well, 34
    as determinant of flow behavior, 24
    as determinant of performance factor, 24
    in diffusivity equation, 31
    equations for determining, 42–43
    increase of, 36
    tests for determining, 41–42, 43–47
Pinchouts
    in buildup test data, 47
    test results used to identify, 3
Pipeline extensions, test results used to design, 3
Pipeline pressure
    as factor in deliverability, 2
    operating, as factor in controlling surface pressure, 2
Porosity
    in analytical solution of diffusivity equation for a single well, 34
    in diffusivity equation, 31–33
Potentials, method of taking gas well, Oklahoma state regulations for, 102–103

Pressure. *See also* Reservoir pressure
    behavior, factors in, 43
    distribution for basic flow system, equation for, 33–34
    drawdown as factor in buildup test, 43
    drop caused by turbulent flow, equation describing, 40
    losses as factor in controlling surface pressure, 2
    losses as factor in deliverability, 3
    measurement, importance of, 79
    measurements, subsurface, sample reporting form for, 207, 208
    measurement in testing, 77–79
    relationships, use of, 28
    surface, controlling, operating pipeline as factor in, 2
Pressure drawdown test. *See* Drawdown test
Pressure survey, annual, single-point test used in, 68
Procedure, rules of. *See* Regulations, state and company
Processing plants, test results used for design of, 3
Production capacity, factors in, 3, 166
Production rates, test results used to determine, 3
Pseudopressure plot, 29–30
Pseudopressure treatment
    calculations for, 25–30
    for gas flow, 24–30
    purpose of, 25
Pseudosteady state flow, 35
    onset of, equation for, 35
    onset of, equation for estimation of, 35
    onset of, in stabilization time, 63
    pressure change during, 36

Quasisteady state flow, 35

Radial diffusivity equation, development of, 31–33
Radial flow, continuity equation for, derivation of, 152–53
Radius of drainage. *See* Drainage radius
Radius of investigation
    defined, 63
    as factor in flow rate duration, 65
    use of, in estimating stabilization time, 63

Test calculation, sample reporting form for, 204

Test, field notes of, sample reporting forms for, 200, 201, 202

Tests
necessity of, 3
purpose of, 2
theoretical fundamentals of, 6–22, 24–52
uses of results of, 3
validity of, 24–25

Test summary, sample reporting form for, 203

Texas, state regulations of, 99–102

Tight gas sands, 18. *See also* Sandface

Tight reservoir(s), tests for, 18

Tortuosity of pores
as determinant of flow behavior, 24
as determinant of performance factor, 24

Transient flow behavior
equation describing, 31–32
equation describing, development of, 31

Transient state flow, 35

Transitional state flow, 35
in reservoir of infinite radius, 36

Turbine meters
advantages of, 72
principle of, 72
use of, 72, 76–77

Turbulence
as determinant of flow behavior, 24
as determinant of performance factor, 24

Turbulent flow, 9, 16
causes of, 40
pressure effects of, 40–41

Turbulent flow factor
equation describing, 40–41
equation describing, unknowns in, 41
equation describing, validity of, 41
friction factor equation for, 57

Unsteady state flow, 35
equation for end of, 35
in reservoir of infinite radius, 36

Viscosity, 8, 10
in analytical solution of diffusivity equation for a single well, 34
calculation of, 180–183
as determinant of flow behavior, 24, 25

as determinant of performance, 10, 24, 25
supercompressibility factor and, determination of, 178–183

Water encroachment and reservoir pressure, 2

Water tables, test results used to identify, 3

Wellbore damage
as determinant of flow behavior, 24
as determinant of performance factor, 24

Wellbore, pressure losses in, as factor in deliverability, 3

Wellbore radius
as determinant of flow behavior, 24
as determinant of performance factor, 24

Wellbore storage
defined, 65
equation for, 65–66
and flow rate duration, 65–66

Zone of altered permeability, 36